Methods in Molecular Biology

For further volumes:
http://www.springer.com/series/7651

For over 35 years, biological scientists have come to rely on the research protocols and methodologies in the critically acclaimed *Methods in Molecular Biology* series. The series was the first to introduce the step-by-step protocols approach that has become the standard in all biomedical protocol publishing. Each protocol is provided in readily-reproducible step-by-step fashion, opening with an introductory overview, a list of the materials and reagents needed to complete the experiment, and followed by a detailed procedure that is supported with a helpful notes section offering tips and tricks of the trade as well as troubleshooting advice. These hallmark features were introduced by series editor Dr. John Walker and constitute the key ingredient in each and every volume of the *Methods in Molecular Biology* series. Tested and trusted, comprehensive and reliable, all protocols from the series are indexed in PubMed.

Thyroid Hormones

Methods and Protocols

Edited by

Constanza Contreras-Jurado

Instituto de Investigaciones Biomédicas Sols-Morreale (IIBM), Consejo Superior de Investigaciones Científicas (CSIC)-Universidad Autónoma de Madrid (UAM), Madrid, Spain; Facultad de Medicina, Universidad Alfonso X El Sabio, Madrid, Spain; Centro de Investigación Biomédica en Red de Cáncer (CIBERONC), Instituto de Salud Carlos III, Madrid, Spain

Humana Press

Editor
Constanza Contreras-Jurado
Instituto de Investigaciones Biomédicas Sols-Morreale (IIBM)
Consejo Superior de Investigaciones Científicas (CSIC)-Universidad
Autónoma de Madrid (UAM)
Madrid, Spain

Facultad de Medicina
Universidad Alfonso X El Sabio
Madrid, Spain

Centro de Investigación Biomédica en Red de Cáncer (CIBERONC)
Instituto de Salud Carlos III
Madrid, Spain

ISSN 1064-3745 ISSN 1940-6029 (electronic)
Methods in Molecular Biology
ISBN 978-1-0716-4251-1 ISBN 978-1-0716-4252-8 (eBook)
https://doi.org/10.1007/978-1-0716-4252-8

This Humana imprint is published by the registered company Springer Science+Business Media, LLC, part of Springer Nature.
The registered company address is: 1 New York Plaza, New York, NY 10004, U.S.A.

Preface

The thyroid hormones (THs), triiodothyronine (T3) and thyroxine (T4), play an essential role in a wide variety of physiological processes such as development, growth, and metabolism. Thyroid hormones (TH) are synthesized in the thyroid gland, whose function is regulated by the hypothalamus-pituitary-thyroid axis and are released into the bloodstream. The classical molecular mechanism of thyroid hormone action is genomic and involves nuclear thyroid hormone receptors (TRs), whereby thyroid hormones enter cells via the ATP-dependent monocarboxylate transporters MCT8 and MCT10 and organic anion transporter proteins. Once inside, thyroid hormones exert their action through nuclear TR receptors that act as ligand-dependent transcription factors. There is also another pathway of action that is not genomic. This pathway begins at a cell surface thyroid hormone receptor analog on the plasma membrane, the αvβ3 integrin, and is very important in promoting cell proliferation in various types of cancer cells.

Therefore, this volume focuses not only on thyroid hormones but also on their transporters and receptors, through a wide range of methods and protocols that are important for advancing the elucidation of the role of thyroid hormones in our body.

This collection covers various aspects of the analysis of the role of thyroid hormones, ranging from conventional methods to state-of-the-art technologies, for in vitro and in vivo studies, through various cellular and experimental animal models and even chapters dedicated to clinical practice in humans.

Overall, this book is likely to appeal to both scientists in fundamental fields and physicians interested in medical applications.

Madrid, Spain ***Constanza Contreras-Jurado***

The original version of the book has been revised. A correction to this book can be found at https://doi.org/10.1007/978-1-0716-4252-8_18

Contents

Contributors

CRISTINA ÁLVAREZ CASTILLA • *Clinic San Carlos Hospital, Madrid, Spain*

ANA ARANDA • *Instituto de Investigaciones Biomédicas "Sols-Morreale", Consejo Superior de Investigaciones Científicas and Universidad Autónoma de Madrid, Madrid, Spain*

MELODY AUTUMN • *Biology Department, Boston College, Chestnut Hill, MA, USA*

SOLEDAD BÁREZ-LÓPEZ • *Laboratory of Thyroid Hormones and Central Nervous System, Department of Neurological Diseases and Aging, Instituto de Investigaciones Biomédicas Sols-Morreale, Consejo Superior de Investigaciones Científicas (CSIC), Universidad Autónoma de Madrid (UAM), Madrid, Spain*

MONTSERRAT CHAO-CRECENTE • *Department of Pathology, Hospital Universitario Severo Ochoa, Leganés, Madrid, Spain; Department of Medicine, Alfonso X El Sabio University, Madrid, Spain*

DARÍO CILLEROS-RODRÍGUEZ • *Departamento de Bioquímica-Instituto de Investigaciones Biomédicas Sols-Morreale (IIBM), Universidad Autónoma de Madrid-Consejo Superior de Investigaciones Científicas, Madrid, Spain*

CONSTANZA CONTRERAS-JURADO • *Instituto de Investigaciones Biomédicas Sols-Morreale (IIBM), Consejo Superior de Investigaciones Científicas (CSIC)-Universidad Autónoma de Madrid (UAM), Madrid, Spain; Facultad de Medicina, Universidad Alfonso X El Sabio, Madrid, Spain; Centro de Investigación Biomédica en Red de Cáncer (CIBERONC), Instituto de Salud Carlos III, Madrid, Spain*

AGUSTÍN F. FERNÁNDEZ • *Nanomaterials and Nanotechnology Research Center (CINN-CSIC), El Entrego, Spain; Health Research Institute of Asturias (ISPA), Oviedo, Spain; University Institute of Oncology (IUOPA), University of Oviedo, Oviedo, Spain; Center for Biomedical Network Research on Rare Diseases (CIBERER), Madrid, Spain*

MARIO F. FRAGA • *Nanomaterials and Nanotechnology Research Center (CINN-CSIC), El Entrego, Spain; Health Research Institute of Asturias (ISPA), Oviedo, Spain; University Institute of Oncology (IUOPA), University of Oviedo, Oviedo, Spain; Center for Biomedical Network Research on Rare Diseases (CIBERER), Madrid, Spain; Department of Organisms and Systems Biology, Area of Plant Physiology, University of Oviedo, Oviedo, Spain*

MARÍA GALLARDO-GÓMEZ • *Translational Oncology Group, Galicia Sur Health Research Institute (IIS Galicia Sur), SERGAS-UVIGO, Vigo, Spain*

CARMEN GRIJOTA-MARTÍNEZ • *Laboratory of Thyroid Hormones and Central Nervous System, Department of Neurological Diseases and Aging, Instituto de Investigaciones Biomédicas Sols-Morreale, Consejo Superior de Investigaciones Científicas (CSIC), Universidad Autónoma de Madrid (UAM), Madrid, Spain*

ANA GUADAÑO-FERRAZ • *Laboratory of Thyroid Hormones and Central Nervous System, Department of Neurological Diseases and Aging, Instituto de Investigaciones Biomédicas Sols-Morreale, Consejo Superior de Investigaciones Científicas (CSIC), Universidad Autónoma de Madrid (UAM), Madrid, Spain*

MARINA GUILLÉN-YUNTA • *Laboratory of Thyroid Hormones and Central Nervous System, Department of Neurological Diseases and Aging, Instituto de Investigaciones Biomédicas*

Sols-Morreale, Consejo Superior de Investigaciones Científicas (CSIC), Universidad Autónoma de Madrid (UAM), Madrid, Spain

TANIA SIERRA HOMPANERA • *Department of Pathology, Hospital Universitario Severo Ochoa, Leganés, Madrid, Spain*

MARINA LASA • *Departamento de Bioquímica-Instituto de Investigaciones Biomédicas Sols-Morreale (IIBM), Universidad Autónoma de Madrid-Consejo Superior de Investigaciones Científicas, Madrid, Spain*

SARAH K. MCMENAMIN • *Biology Department, Boston College, Chestnut Hill, MA, USA*

NORBERTO HERRERA MERINO • *Department of Medicine, Alfonso X El Sabio University, Madrid, Spain*

ANA MONTERO-PEDRAZUELA • *Laboratory of Thyroid Hormones and Central Nervous System, Department of Neurological Diseases and Aging, Instituto de Investigaciones Biomédicas Sols-Morreale, Consejo Superior de Investigaciones Científicas (CSIC), Universidad Autónoma de Madrid (UAM), Madrid, Spain; Instituto de Investigaciones Biomédicas Sols-Morreale, Consejo Superior de Investigaciones Científicas (CSIC), Universidad Autónoma de Madrid (UAM), Madrid, Spain*

RAÚL F. PÉREZ • *Nanomaterials and Nanotechnology Research Center (CINN-CSIC), El Entrego, Spain; Health Research Institute of Asturias (ISPA), Oviedo, Spain; University Institute of Oncology (IUOPA), University of Oviedo, Oviedo, Spain; Center for Biomedical Network Research on Rare Diseases (CIBERER), Madrid, Spain; Institut d'Investigacions Biomèdiques August Pi i Sunyer (IDIBAPS), Barcelona, Spain; Department of Biochemistry and Molecular Biology, Faculty of Veterinary Medicine, Complutense University of Madrid, Madrid, Spain*

ISABELLA RANIERI • *Biology Department, Boston College, Chestnut Hill, MA, USA*

DIEGO RODRÍGUEZ MUÑOZ • *Clinical Research and Regulatory Associate, MDx, Madrid, Spain*

CRISTINA SAIZ-LADERA • *Unidad de Oncogenómica, Servicio de Oncohematología, Fundación para la Investigación Biomédica del Hospital Universitario Niño Jesús, Madrid, Spain*

ÁNGELA SÁNCHEZ • *Department of Metabolic and Immune Diseases, Instituto de Investigaciones Biomédicas Sols-Morreale Centro Mixto Consejo Superior de Investigaciones Científicas CSIC-Universidad Autónoma de Madrid, Madrid, Spain*

PABLO SANTAMARINA-OJEDA • *Nanomaterials and Nanotechnology Research Center (CINN-CSIC), El Entrego, Spain; Health Research Institute of Asturias (ISPA), Oviedo, Spain; University Institute of Oncology (IUOPA), University of Oviedo, Oviedo, Spain; Center for Biomedical Network Research on Rare Diseases (CIBERER), Madrid, Spain*

VÍCTOR VALCÁRCEL-HERNÁNDEZ • *Laboratory of Thyroid Hormones and Central Nervous System, Department of Neurological Diseases and Aging, Instituto de Investigaciones Biomé dicas Sols-Morreale, Consejo Superior de Investigaciones Científicas (CSIC), Universidad Autónoma de Madrid (UAM), Madrid, Spain; Laboratory of Molecular Physiology and Adaptation, CNRS UMR7221, Department of Adaptations of Life, Muséum National d'Histoire Naturelle, Paris, France*

NICOLE DIMAS VASQUEZ • *Department of Pathology, Hospital Universitario Severo Ochoa, Leganés, Madrid, Spain*

MARÍA VICTORIA REDONDO VEGA • *Department of Medicine, Alfonso X El Sabio University, Madrid, Spain*

JENNY ZENG • *Biology Department, Boston College, Chestnut Hill, MA, USA*

Part I

An Overview

Chapter 1

Thyroid Hormones and Co-workers: An Overview

Constanza Contreras-Jurado

Abstract

The hypothalamus secretes the thyroid-releasing hormone (TRH) that induces the pituitary gland to release the thyroid-stimulating hormone (TSH) which stimulates thyroid follicular cells to release the thyroid hormones (THs), thyroxine (T4), and triiodothyronine (T3). The process of synthesizing T3 and T4 hormones involves various enzymatic steps, starting with the iodination of L-tyrosine residues present in the protein thyroglobulin. Thyroid hormones are released into the bloodstream, where they bind to thyroid hormone distributor proteins (THDPs) which transport them in the circulation. The conversion of T4 to T3 (the more biologically active hormone) in target tissues is facilitated by selenoprotein enzymes known as deiodinases. THs can bind to different molecules located on the plasma membrane, such as integrin αvβ3, through which they exercise regulatory non-genomic control. Nevertheless, most of thyroid hormone's actions are mediated intracellularly by binding to thyroid hormone receptors (TRs). Thyroid hormone receptors act as ligand-dependent transcription factors, Thyroid hormone receptors activate thyroid hormone response elements on gene promoters through canonical signaling. Thyroid hormones mediate several critical physiological processes including organ development, cell differentiation, metabolism, and cell growth and maintenance.

Key words Thyroid hormones (THs), Thyroid hormone receptors (TRs), Integrin αvβ3, Thyroid hormone transporter, Hypothyroidism, Hyperthyroidism, Deiodinases

1 The Hypothalamus-Pituitary-Thyroid Axis

The hypothalamus and the pituitary gland are in close anatomical proximity at the base of the brain and extend through the pituitary stalk to the sella turcica. The target organs are located peripherally and function through stimulation or inhibition by pituitary hormones.

The hypothalamus secretes the thyroid-releasing hormone (TRH) that induces the pituitary gland to release the thyroid-stimulating hormone (TSH) which stimulates thyroid follicular cells to release the thyroid hormones (THs), thyroxine (T4) (80%), and triiodothyronine (T3) (20%). The secretion of TRH and TSH is controlled by negative feedback from THs to

Constanza Contreras-Jurado (ed.), *Thyroid Hormones: Methods and Protocols*, Methods in Molecular Biology, vol. 2876,
https://doi.org/10.1007/978-1-0716-4252-8_1,

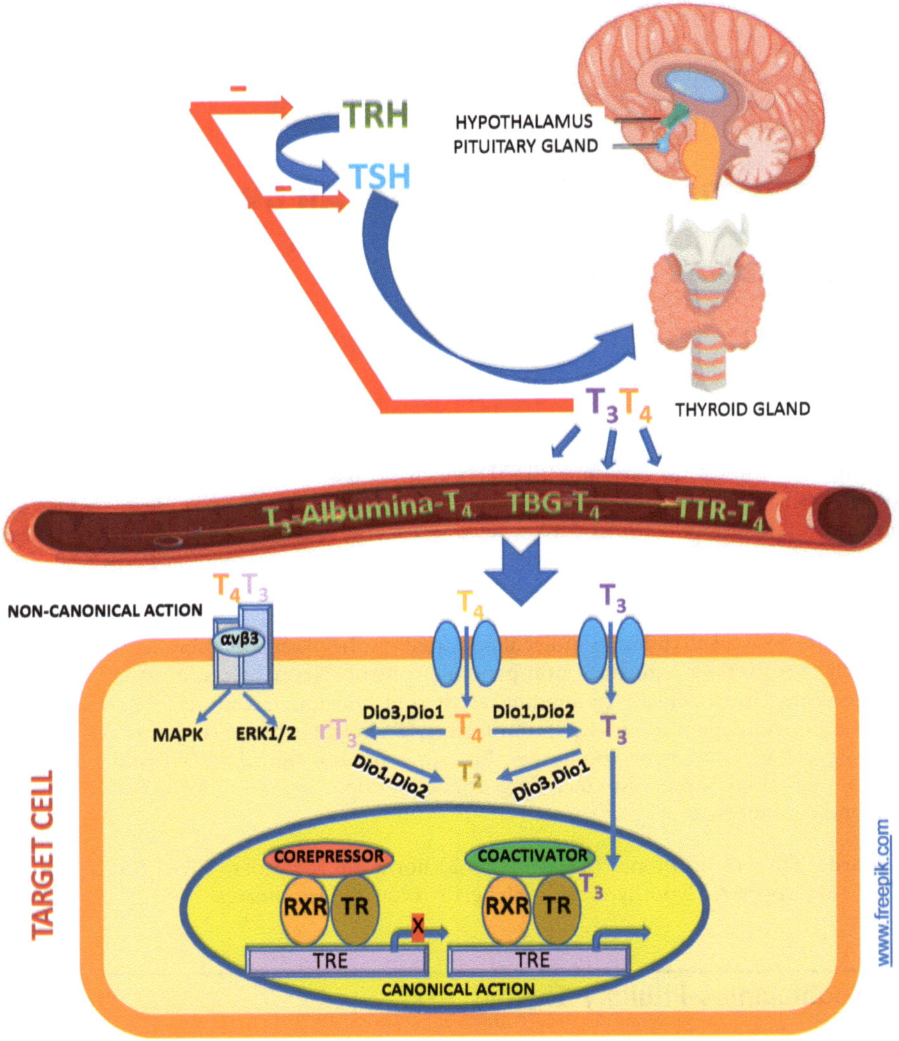

Fig. 1 The hypothalamus secretes TRH that induces the pituitary gland to release TSH which stimulates thyroid follicular cells to release T4 and T3. THs are released into the bloodstream, where they bind to thyroid hormone distributor proteins. The conversion of T4 to T3 in target tissues is facilitated by deiodinases. The classical genomic actions (canonical action) carried out by T3 involve the mediation of TR, which bind TREs and directly regulate gene expression. The non-genomic actions (non-canonical action) of THs are mediated by integrin αvβ3 and exert a regulatory influence on a variety of genes, like MAPK and ERK1/2

maintain physiological levels of the main hormones of the hypothalamus-pituitary-thyroid (HPT) axis [1] (Fig. 1).

2 Thyroid Gland and Thyroid Hormones

THs, specifically T3 and T4, are synthesized and secreted by the thyroid gland, an endocrine organ centrally located in the anterior-lower region of the neck, positioned between the fifth cervical vertebra and the first thoracic vertebra. Comprising two lobes, the right and left, that are united by an intermediary structure known as the isthmus. Within the thyroid lobules, numerous functional units called thyroid follicles facilitate the secretion of thyroid hormones by specialized cells. The process of synthesizing T3 and T4 hormones involves various enzymatic steps, starting with the iodination of L-tyrosine residues present in the protein thyroglobulin (TG). This iodination takes place at position 3 or both positions 3 and 5 of the aromatic ring. After that, the mono-iodinated (MIT) or di-iodinated (DIT) ring of a tyrosine residue is transferred to another di-iodinated tyrosine ring, leading to the formation of triiodothyronine (T3) or thyroxine (T4) residues, respectively. Both T3 and T4 contain an inner tyrosine ring and an outer phenolic ring, and they are the immediate precursors of T3 and T4 hormones. The synthetic process is catalyzed by a thyroid peroxidase (TPO) [2]. The predominant circulating form of thyroid hormone is T4, accompanied by a smaller quantity of T3. It is noteworthy that, while the initial secretion is primarily in the form of T4, the conversion to the more biologically active T3 occurs subsequently. This conversion underscores the dynamic nature of thyroid hormone regulation within the body [3]. THs are released into the bloodstream, where they bind to TH distributor proteins (THDPs) which transport them in the circulation. There are two main TH carriers in the blood: thyroxine-binding globulin (TBG), a monomeric protein of 54 kDa that shows higher affinity for THs, and transthyretin (TTR), a symmetrical tetramer of 55 kDa, rich in beta-sheet structures. Additionally, albumin can also bind THs, albeit with lower affinity than TBG and TTR. These carriers have a dual function: They control the amount of free hormone exchangeable with the cells and are responsible for the uniform distribution of circulating THs throughout the body. THs and some of their metabolites can also bind to low-density lipoproteins (LDL).

The conversion of T4 to T3 in target tissues is facilitated by selenoprotein enzymes known as deiodinases [4]. Deiodinase enzymes form a group of dimeric integral membrane proteins with a thioredoxin fold, capable of modulating thyroid hormone activity. Their function is dichotomous, as they can either activate or deactivate thyroid hormone, contingent upon whether they target the phenolic or tyrosyl rings of the iodothyronines, respectively. Specifically, type I and II deiodinases (DIO1 and DIO2) play a crucial role in converting T4 to T3 in different tissues, thereby elevating the levels of circulating T3. Conversely, the inner ring DIO3 enzyme is responsible for hormone inactivation, as it converts T4 and T3 into inactive metabolites, namely reverse T3 (rT3) or T2, respectively. This regulatory mechanism ensures a finely tuned balance between active and inactive forms of thyroid hormones, influencing their impact on various tissues and physiological processes. These three deiodinases have distinct tissue distributions, catalytic specificities, physiological functions, and regulations [5].

The thyroid hormone signaling pathway is an intricate and meticulously regulated process due to the presence of cell and tissue-specific thyroid hormone transporters, plasma membrane receptors, multitude of thyroid hormone receptor (TR) isoforms, and intricate interactions with corepressors and coactivators. Furthermore, thyroid signals frequently engage in crosstalk with various other signaling pathways.

Thyroid hormones, specifically T3 and T4, stand as indispensable regulatory molecules in the human body, playing pivotal roles in growth, development, metabolism, and immune system function [6]. Imbalances, whether in deficiency or excess of THs, are linked to severe disorders affecting diverse organs. Although THs can bind to different molecules located on the plasma membrane, such as integrin αvβ3 [7], most of their actions are mediated intracellularly by TH transporter proteins such as monocarboxylate transporter (MCT)8, MCT10, and organic anion-transporting polypeptide 1C1, and by binding to TRs. T3 has a higher affinity than T4 for TRs, while T4 is more potent than T3 in binding integrin αvβ3 [8].

3 Genomic and Non-genomic Actions of Thyroid Hormones

The classical genomic actions carried out by T3 involve the mediation through high-affinity nuclear receptors (NRs), which bind thyroid hormone response elements (TREs) and directly regulate gene expression. In contrast, T4 has a higher affinity to membrane receptors like integrin αvβ3 through which it exercises regulatory non-genomic control. These effects manifest swiftly and remain unaffected by inhibitors of transcription and protein synthesis. The non-genomic actions of THs are thought to be mediated by induction of [Ca2+], cyclic-AMP, or the protein kinase signaling cascades. Nongenomic actions of thyroid hormone typically occur quickly, within seconds to minutes. For these actions to take place, the hormones need to have swift access to the tissue target of the action. The epithelial cells of the thyroid gland (thyrocytes) can concentrate iodide ions thanks to a specific sodium/iodide symporter (NIS or SLC5), which is present in their basolateral membrane. The synthesis of this symporter is stimulated by TSH. In contrast, T3-dependent transcriptional actions occur over hours to days [9].

3.1 Plasma Membrane Integrin $\alpha v \beta 3$

Integrins constitute a distinctive category of cell adhesion molecules that play a crucial role in linking the cytoskeleton to the extracellular matrix (ECM) or neighboring cells. Comprising two genetically unrelated subunits, α and β, these molecules form non-covalent associations to facilitate their functional integration. The specific integrin αvβ3, identified as the vitronectin receptor, is composed of a 125-kDa αv subunit and a 105-kDa β3 subunit, stands as a heterodimeric structural protein embedded in the plasma membrane, serving as a cell surface receptor specifically dedicated to thyroid hormone recognition. Its role diverges significantly from that of the conventional TR designed for thyroid hormone signaling.

Thyroid hormone, within the extracellular domain of αvβ3, exerts a regulatory influence on a variety of genes, like the activation of extracellular signal-regulated kinases ERK1/2, PI3K, contributing to the modulation of processes such as angiogenesis, cancer cell proliferation, metastasis, and defense pathways in cancer cells. Another nongenomic role for TH is the phosphorylation of proteins. It not only stabilizes and facilitates protein phosphorylation within synaptosomes and intermediate filaments but also exerts its influence on the mature and developing cytoskeleton in the cerebral cortex. Furthermore, T3, a specific form of thyroid hormone, induces alterations in the phosphorylation status of various kinases, such as p38. This multifaceted impact underscores the intricate regulatory role that TH and T3 play in modulating cellular processes, particularly in the dynamic environment of the cerebral cortex [6].

The hormone, itself, does not enter the cell as a part of these functions. After T4 interaction, integrin αvβ3 is endocytosed into the cytoplasm. Monomeric αv translocates to the nucleus and may function as a co-activator protein [10].

3.2 TH Transporter Proteins

THs predominantly exert their effects through binding to specific nuclear thyroid receptors. Transmembrane transporter proteins facilitate the entry of free THs into the cells, rather than passive crossing of plasma membranes as previously believed. TH uptake is saturable, energy-dependent, and Na+-dependent in certain cell types. This observation led to the conclusion that transporter proteins facilitate the transport of TH across the plasma membrane. The first TH transporter identified was the organic anion transporter subtype (oatp3) [11]. Among the diverse thyroid hormone transporters, encompassing up to 16 members across five distinct protein families, MCT8, MCT10, organic anion transporting polypeptide (OATP) 1C1, and SLC17A4 exhibit the highest specificity for thyroid hormones. Notably, MCT8 stands out as the most specific thyroid hormone transporter identified to date. Located on the X chromosome, MCT8 is predominantly expressed in the hypothalamus. Its deficiency, as observed in Allan-Herndon-

Dudley syndrome, manifests as a neurocognitive (central) impairment linked to a hypothyroid state in the brain, alongside a peripheral manifestation attributed to T3 excess in the circulation. Current exploratory treatment strategies for MCT8 deficiency include thyroid hormone analog therapy, molecular chaperones, and gene therapy, reflecting the ongoing efforts to address this intricate physiological imbalance [12].

3.3 TH Nuclear Receptors (TRs)

T3 stands out as the most potent TH, actively participating in the mediation of gene expression by binding to TRs found either in the cytoplasm or nucleus of the cell. TRs belong to the nuclear receptor superfamily, which includes receptors for steroid hormones, retinoids, vitamin D, and various "orphan" receptors. Acting as ligand-dependent transcription factors, TRs activate TREs on gene promoters through canonical signaling. TRs exert pivotal physiological roles in the regulation of developmental processes, growth, metabolism, and immune system modulation [13].

Encoded by two distinct genes (α and β) located on human chromosomes 17 and 3, respectively, TR isoforms exhibit highly homologous amino acid sequences and distinct expression patterns during development and in adult tissues. The primary hormone-binding isoforms include TRα1, TRβ1, and TRβ2. TRα1, predominantly expressed in the brain, heart, and skeletal muscle, has one T3-binding splice product, while TRα2 and TRα3 are non–T3-binding splice products, along with additional truncated forms. TRβ1 is widely expressed, TRβ2 is primarily found in the brain, retina, and inner ear, and TRβ3 is expressed in the kidney, liver, and lung [14].

The thyroid hormone receptors demonstrate remarkable dynamism, with TRα1 and TRβ1 dynamically shuttling between the nucleus and cytoplasm. The biological effects of thyroid hormones in a given tissue depend on factors such as hormone availability, levels of different TR subtypes, posttranslational modifications, heterodimerization partners, and interactions with corepressors and coactivators. The multifaceted cascade of events involves TR binding to TREs, leading to the modulation of target gene expression in response to thyroid hormone.

Notably, thyroid hormone receptors also exhibit cytoplasmic functions, as certain TR subtypes have been identified in the mitochondria, plasma membrane, and cytoplasmic compartments in a tissue-specific manner. This diversification of subcellular localization further underscores the complexity of thyroid hormone receptor functions [15].

3.4 Coactivators and Corepressors

NR coregulators are molecules that play a crucial role in the efficient functioning of nuclear receptors. NR coregulators are grouped into two categories, namely nuclear receptor coactivators (NCOAs) and corepressors (NCORs), based on their ability to activate or repress gene transcription, respectively [16]. However, their functions are highly context-dependent, and the direction of transcriptional regulation is not an intrinsic feature of the coregulators. Therefore, the names NCOAs and NCORs do not accurately reflect their functions [17]. When a molecule binds to a receptor on a cell, it triggers the recruitment of coactivator complexes, which then increase the level of histone acetylation. This, in turn, attracts RNA polymerase II (Pol II) to activate the transcription process. In the case of thyroid hormone-regulated target genes, different patterns of coactivator recruitment and histone acetylation are observed. This phenomenon may enable highly specific regulation of gene expression [18]. Genetic variants of NCOAs or NCORs have started emerging in humans with obesity, hormonal disturbance, intellectual disability, and autism spectrum disorders [19, 20]. On positive TREs, corepressors such as N-CoR1 or N-CoR2 (also known as SMRT) and histone deacetylase (HDAC) intricately associate with TRs in the absence of ligands, resulting in the suppression of target gene expression [21, 22]. Upon the binding of ligands, a dynamic conformational shift occurs in the TR, facilitating the engagement of a distinct set of activator proteins with the receptor, including SRC-1 (p160/steroid receptor coactivator (1) and histone acetyltransferase (HAT). This molecular rearrangement induces alterations in chromatin structure, subsequently promoting the transcriptional activation of the target gene. Beyond its role in activating transcription on positive TREs, TRs can also exert a repressive influence on gene expression, potentially by binding to putative negative TREs in a T3-dependent manner. The involvement of N-CoR1 and SMRT in determining T3 sensitivity implies that these corepressors play a pivotal role in recruiting the TR in the presence of T3, adding a layer of complexity to the regulatory mechanisms governing thyroid hormone signaling [23, 24].

4 Clinical Implications

4.1 Thyroid Cancer

Thyroid cancer is the most malignant endocrine neoplasm and although relatively rare, the development of thyroid nodules is a very common phenomenon. It is estimated that 50% of the general population has nodules detectable by ultrasound. Of the total number of nodules analyzed only 5% of them are malignant [25]. There are three major histological types of thyroid cancer: differentiated thyroid cancers (DTCs), which derive from the epithelial cells of the thyroid follicles; medullary thyroid cancers (MTCs), which begin in the calcitonin-secreting "C" cells; and anaplastic cancers (CATs). Differentiated thyroid carcinomas account for 90% of thyroid tumors diagnosed, with papillary thyroid carcinoma (PTC) being the most common type, constituting approximately 80% of all thyroid cancers and a survival rate of 95%. Follicular thyroid carcinoma (FTC) accounts for 10% to 15% of all thyroid cancers with a 91% survival rate. CMT is a rare variety of thyroid cancer that accounts for 5% of all thyroid cancers with a survival rate of 89%. CAT is a very rare thyroid cancer and accounts for only 2–5% of all thyroid cancers, it usually occurs at advanced ages, it is the most aggressive as it metastasizes distantly, and its prognosis is poor with a survival rate of only 1–7% [26]. CAT is a very rare thyroid cancer and accounts for only 2–5% of all thyroid cancers. The main diagnostic test for the study of thyroid nodules is cervical ultrasound, although it only provides information regarding the presence or absence of nodules and information regarding glandular size and echogenicity. Thyroid nodules can show a very diverse ultrasound pattern which often makes it difficult to classify them as malignant. This test uses the classification known as TIRADS (Thyroid Imaging Reporting and Data System) although it does not provide a suspected diagnosis of malignancy [27]. Nodules with a TIRADS 3 and TIRADS 4 classification although mildly suspicious with a malignancy in less than 5% of cases and moderately suspicious with a risk of malignancy varying between 10 and 80% of cases, respectively, must be studied in more detail by fine needle aspiration-puncture (FNA) to obtain cytologic material. The puncture results in material to be analyzed by the anatomopathologist who will give the diagnosis on a risk scale (the BETHESDA scale). This diagnostic test is useful in the case of papillary, medullary, and anaplastic tumors but it is impossible to distinguish cytologically between adenoma and follicular carcinoma because histological evidence of capsular and vascular invasion is required, which is not defined with FNA, Therefore, a high percentage of patients with tumors classified as Bethesda III and IV whose cytologic result is indeterminate have to undergo hemithyroidectomy (partial or total resection of the thyroid gland) to histologically analyze the nodule even though only 5% of all thyroid nodules are malignant [28]. This leads to subjecting the patient to a series of possible complications (hypothyroidism, paralysis of the vocal cords, cervical hematoma which is very dangerous, transient, or permanent hypocalcemia which can lead to tetany, changes in the electrocardiogram, and even convulsions) and to a health care cost that would not be necessary if reliable markers were available to differentiate benign from malignant thyroid nodules [29].

4.2 Hypothyroidism

Hypothyroidism is a common condition caused by a deficiency of thyroid hormones. If left untreated, it can lead to serious health problems and even death. When thyroid-stimulating hormone (TSH) levels are above the reference range and free thyroxine levels are below the reference range, it's known as overt or clinical primary hypothyroidism. Subclinical hypothyroidism is when TSH levels are above the reference range, but free thyroxine levels are within the normal range. Hypothyroidism can also result from severe iodine deficiency because thyroid hormone synthesis requires iodine. Hashimoto thyroiditis is a chronic autoimmune thyroid disease and is the most common cause of primary hypothyroidism in areas with sufficient iodine. Since thyroid hormone is crucial to normal development in childhood, most developed countries have established neonatal screening programs to identify congenital hypothyroidism. The prevalence of congenital hypothyroidism is between 1 in 500 to 3000 newborns depending on ethnicity [30]. Additionally, these countries have implemented programs to prevent severe iodine deficiency, such as universal salt iodization. Another common cause of hypothyroidism is total thyroidectomy, a surgical procedure removing the entire thyroid gland, radioiodine therapy, employed in the treatment of hyperthyroidism, and medications such as immune checkpoint inhibitors and tyrosine kinase inhibitors [31, 32].

Levothyroxine, a synthetic form of T_4, is the standard treatment for hypothyroidism. However, some patients continue to experience symptoms even after reaching the biochemical therapy targets. This has led to the question of whether levothyroxine treatment is sufficient for all patients or whether alternative therapies, like combination with liothyronine preparations, should be considered [33].

Hypothyroidism is more common in women, older individuals, and white people. It's also more prevalent in patients with autoimmune diseases like type 1 diabetes, autoimmune gastric atrophy, and coeliac disease [34]. Hypothyroidism can be classified as primary (due to thyroid hormone deficiency), secondary (due to TSH deficiency), tertiary (due to thyrotropin-releasing hormone deficiency), and peripheral (extra-thyroidal). Central hypothyroidism (including both secondary and tertiary) and peripheral hypothyroidism are rare and account for less than 1% of cases [35]. The most common symptoms of hypothyroidism in adults are fatigue, lethargy, cold intolerance, weight gain, constipation, change in voice, and dry skin. However, the presentation of symptoms can vary widely depending on age, gender, and time between onset and diagnosis. Patients with hypothyroidism have a higher prevalence of cardiovascular risk factors and often have features of metabolic syndrome, including hypertension, increased waist circumference, and dyslipidemia. Other symptoms include neurosensory, musculoskeletal, and gastrointestinal signs and symptoms [36].

4.2.1 Clinical Presentation and Significance

(i).

Myxedema coma represents a rare but critical medical emergency characterized by an array of symptoms, including altered mental status, hypothermia, progressive lethargy, and bradycardia. Left untreated, it can progress to multiple organ dysfunction syndrome and, ultimately, mortality. Thus, the prompt initiation of thyroid hormone therapy and other essential supportive measures is paramount in managing this condition [37].

(ii).

Severe primary hypothyroidism poses additional complexities, potentially leading to pituitary hyperplasia and concurrent pituitary pathology, such as secondary adrenal insufficiency, with associated symptoms like amenorrhea [38].

In adults, hypothyroidism commonly manifests with symptoms such as fatigue, lethargy, cold intolerance, weight gain, constipation, voice changes, and dry skin. However, the clinical presentation can vary widely based on factors such as age, sex, and the time between onset and diagnosis. Beyond the classic symptoms, hypothyroidism has far-reaching clinical implications, impacting nearly all major organ systems. Patients with hypothyroidism exhibit a higher prevalence of cardiovascular risk factors and frequently present features of metabolic syndrome, including hypertension, increased waist circumference, and dyslipidemia. The manifestations extend to neurosensory, musculoskeletal, and gastrointestinal signs and symptoms, emphasizing the diverse impact of hypothyroidism on overall health. Recognizing and addressing these varied presentations is crucial for comprehensive patient management and improving outcomes [36].

4.3 Hyperthyroidism

Hyperthyroidism is a condition where an excessive amount of thyroid hormones is present in the body due to increased synthesis of thyroid hormones, excessive release of preformed thyroid hormones, or an endogenous or exogenous extrathyroidal source. The most common causes of excessive production of thyroid hormones are Graves' disease, toxic multinodular goiter, and toxic adenoma. The most common cause of excessive passive release of thyroid hormones is painless (silent) thyroiditis, which has the same clinical presentation as other causes.

Hyperthyroidism caused by overproduction of thyroid hormones can be treated with antithyroid medications such as methimazole and propylthiouracil, radioactive iodine ablation of the thyroid gland, or surgical thyroidectomy.

Nonautoimmune hyperthyroidism or hereditary hyperthyroidism is a rare form of hyperthyroidism characterized by excessive thyroid activity. Major symptoms include hyperactivity, anxiety, weight loss, exophthalmos, and tachycardia. Clinical diagnosis is based on observation and measurement of plasma concentrations of thyroid hormone. Differential diagnosis is based on the absence of exophthalmos and the presence of myxedema, anti-TSH antibodies, and lymphocyte infiltration of the thyroid [39].

4.4 Resistance to Thyroid Hormone

Thyroid hormone resistance syndromes refer to a group of genetic conditions characterized by decreased sensitivity of body tissues to thyroid hormones. The incidence of thyroid hormone resistance syndromes is probably 1 case per 50,000 live births. This condition is characterized by high levels of serum-free thyroid hormones, and normal or elevated TSH levels.

4.4.1 Mutations in TRα and TRβ Genes

In most cases, thyroid hormone resistance is caused by mutations in the TRβ gene but also been describing mutations in the TRα gene. Resistance to TRβ has elevated circulating T4 and T3 with non-suppressed pituitary TSH levels reflecting central (hypothalamic–pituitary) refractoriness to thyroid hormone action and is associated with variable resistance to hormone action in peripheral tissues. Patients diagnosed with RTH-β exhibit a varied phenotype which includes symptoms such as goiter, tachycardia, raised energy expenditure, hyperactive behavior, delayed bone age, and learning disabilities. These symptoms are caused by a combination of low action of TH in predominantly TRβ-expressing tissues, and overexposure of TH in TRα-expressing tissues. Mutant TRß proteins have a reduced affinity for T3, and/or abnormal interaction with cofactors, such as decreased interaction with co-activators, or increased interaction with co-repressors.

In thyroid hormone resistance caused by mutations in the TRα was not expected significant changes in serum TH levels as TRα1 is not involved in the negative feedback action of TH, but the phenotype observed in these patients includes growth retardation, delayed bone development, mildly delayed motor and mental development, abnormal thyroid function tests, and low GH and IGF1 levels [40].

4.4.2 Deiodination Defect

Metabolism of TH is importantly controlled by the iodothyronine deiodinases. Mutations in the SBP2 protein (also known as Sec insertion sequence-binding protein 2 SECISBP2), which is required for normal deiodination, give rise to a multisystem disorder including abnormal thyroid function tests. This syndrome has a circulating thyroid hormone profile consisting of elevated T4 and rT3 and decreased T3. In all identified families, the most prominent features include growth retardation, mental and motor retardation, muscle weakness, hypoglycemia, and impaired hearing. Infertility is variably reported. These conditions are associated with the reduced synthesis of most of the 25 known human selenoproteins, resulting in a complex phenotype [41].

4.4.3 Relevance of mct8 in Health and Disease

While numerous proteins play roles in thyroid hormone transport, mutations in genes encoding MCT8, MCT10, and OATP1C1 have emerged as pivotal factors with direct pathophysiological consequences. Of these, MCT8 stands out as the sole transporter specifically designed for the conveyance of thyroid hormones and certain derivatives. This transmembrane transporter is expressed at the blood–brain barrier (BBB), and in neural cells. Deficiencies in MCT8 disrupt the efficient transfer of thyroid hormones across the blood–brain barrier.

An in-depth exploration of SLC16A2, the gene responsible for encoding MCT8 reveals that mutations in this gene result in an X-linked syndrome characterized by severe neurological impairment and discernible alterations in thyroid hormone concentrations. Notably, individuals with compromised MCT8 function exhibit prenatal onset of brain lesions, strongly implicating cerebral hypothyroidism as a contributory factor.

Furthermore, MCT8 mutations manifest in the clinical presentation of Allan–Herndon–Dudley syndrome, marked by truncal hypotonia. This syndrome progresses to spastic quadriplegia, speech impairment, severe intellectual deficits, and disruptions in thyroid hormone concentrations. The intricate interplay between MCT8 mutations and the intricate cascade of physiological consequences underscores the critical role of this transporter in both health and disease contexts [42, 43].

References

1. Ortiga-Carvalho TM, Chiamolera MI, Pazos-Moura CC et al (2016) Hypothalamus-pituitary-thyroid Axis. Compr Physiol 6(3): 1387–1428. https://doi.org/10.1002/cphy.c150027
2. Giammanco M, Di Liegro CM, Schiera G et al (2020) Genomic and non-genomic mechanisms of action of thyroid hormones and their catabolite 3,5-Diiodo-L-Thyronine in mammals. Int J Mol Sci 21(11):4140. https://doi.org/10.3390/ijms21114140
3. Chaudhary P, Singh Z, Khullar M et al (2013) Levator glandulae thyroideae, a fibromusculo-glandular band with absence of pyramidal lobe and its innervation: a case report. J Clin Diagn Res 7(7):1421–1424. https://doi.org/10.7860/JCDR/2013/6144.3186
4. Kaplan MM, Breitbart R (1984) Conversion of thyroxine to triiodothyronine in the anterior pituitary gland and the influence of this process on thyroid status. Horm Metab Res Suppl 14: 79–85
5. Bianco AC, Kim BW (2006) Deiodinases: implications of the local control of thyroid hormone action. J Clin Invest 116(10): 2571–2579. https://doi.org/10.1172/JCI29812
6. Mullur R, Liu YY, Brent GA (2014) Thyroid hormone regulation of metabolism. Physiol Rev 94(2):355–382. https://doi.org/10.1152/physrev.00030.2013
7. Bergh JJ, Lin HY, Lansing L et al (2005) Integrin alphaVbeta3 contains a cell surface receptor site for thyroid hormone that is linked to activation of mitogen-activated protein kinase and induction of angiogenesis. Endocrinology 146(7):2864–2871. https://doi.org/10.1210/en.2005-0102
8. Tedeschi L, Vassalle C, Iervasi G et al (2021) Main factors involved in thyroid hormone action. Molecules 26(23):7337. https://doi.org/10.3390/molecules26237337
9. Davis PJ, Davis FB (1996) Nongenomic actions of thyroid hormone. Thyroid 6(5): 497–504. https://doi.org/10.1089/thy.1996.6.497
10. Lin HY, Cody V, Davis FB, Hercbergs AA, Luidens MK, Mousa SA, Davis PJ (2011) Identification and functions of the plasma membrane receptor for thyroid hormone analogues. Discov Med 11(59): 337–347. PMID: 21524387
11. Abe T, Kakyo M, Sakagami H et al (1998) Molecular characterization and tissue distribution of a new organic anion transporter subtype (oatp3) that transports thyroid hormones and taurocholate and comparison with oatp2. J Biol Chem 273(35):22395–22401. https://doi.org/10.1074/jbc.273.35.22395

12. Groeneweg S, van Geest FS, Peeters RP et al (2020) Thyroid hormone transporters. Endocr Rev 41(2):bnz008. https://doi.org/10.1210/endrev/bnz008
13. Weikum ER, Liu X, Ortlund EA (2018) The nuclear receptor superfamily: a structural perspective. Protein Sci 27(11):1876–1892. https://doi.org/10.1002/pro.3496
14. Aranda A, Alonso-Merino E, Zambrano A (2013) Receptors of thyroid hormones. Pediatr Endocrinol Rev 11(1):2–13
15. Anyetei-Anum CS, Roggero VR, Allison LA (2018) Thyroid hormone receptor localization in target tissues. J Endocrinol 237(1):R19–R34. https://doi.org/10.1530/JOE-17-0708
16. Lonard DM, Lanz RB, O'Malley BW (2007) Nuclear receptor coregulators and human disease. Endocr Rev 28(5):575–587. https://doi.org/10.1210/er.2007-0012
17. O'Malley BW, McKenna NJ (2008) Coactivators and corepressors: what's in a name? Mol Endocrinol 22(10):2213–2214. https://doi.org/10.1210/me.2008-0201
18. Liu Y, Xia X, Fondell JD et al (2006) Thyroid hormone-regulated target genes have distinct patterns of coactivator recruitment and histone acetylation. Mol Endocrinol 20(3):483–490. https://doi.org/10.1210/me.2005-0101
19. Dasgupta S, Lonard DM, O'Malley BW (2014) Nuclear receptor coactivators: master regulators of human health and disease. Annu Rev Med 65:279–292. https://doi.org/10.1146/annurev-med-051812-145316
20. Rollins DA, Coppo M, Rogatsky I (2015) Minireview: nuclear receptor coregulators of the p160 family: insights into inflammation and metabolism. Mol Endocrinol 29(4):502–517. https://doi.org/10.1210/me.2015-1005
21. Oberoi J, Fairall L, Watson PJ et al (2011) Structural basis for the assembly of the SMRT/NCoR core transcriptional repression machinery. Nat Struct Mol Biol 18(2):177–184. https://doi.org/10.1038/nsmb.1983
22. Mendoza A, Astapova I, Shimizu H et al (2017) NCoR1-independent mechanism plays a role in the action of the unliganded thyroid hormone receptor. Proc Natl Acad Sci USA 114(40):E8458–E8467. https://doi.org/10.1073/pnas.1706917114
23. Shimizu H, Astapova I, Ye F et al (2015) NCoR1 and SMRT play unique roles in thyroid hormone action in vivo. Mol Cell Biol 35(3):555–565. https://doi.org/10.1128/MCB.01208-14
24. Astapova I, Vella KR, Ramadoss P et al (2011) The nuclear receptor corepressor (NCoR) controls thyroid hormone sensitivity and the set point of the hypothalamic-pituitary-thyroid axis. Mol Endocrinol 25(2):212–224. https://doi.org/10.1210/me.2010-0462
25. Kitahara CM (2016) The changing incidence of thyroid cancer. Nat Rev Endocrinol 12:646–653
26. Noone HN, et al. SEER cancer statistics review, 1975–2017, National Cancer Institute, Bethesda. https://seer.cancer.gov/csr/1975_2017/, based on November 2019 SEER data submission, posted to the SEER website, April 2020
27. Hovarth E (2009) An ultrasonogram reporting system for thyroid nodules stratifying cancer risk for clinical management. J Clin Endocrinol Metab 94:1748–1751
28. Patel KN et al (2020) The American Association of Endocrine Surgeons Guidelines for the definitive surgical Management of Thyroid Disease in adults. Ann Surg 271(3):e21–e93
29. Chisholm EJ, Kulinskaya E, Tolley NS (2009) Systematic review and meta-analysis of the adverse effects of thyroidectomy combined with central neck dissection as compared with thyroidectomy alone. Laryngoscope 119
30. Deladoëy J, Ruel J, Giguère Y et al (2011) Is the incidence of congenital hypothyroidism really increasing? A 20-year retrospective population-based study in Québec. J Clin Endocrinol Metab 96(8):2422–2429. https://doi.org/10.1210/jc.2011-1073
31. Fallahi P, Ferrari SM, Vita R et al (2014) Thyroid dysfunctions induced by tyrosine kinase inhibitors. Expert Opin Drug Saf 13(6):723–733. https://doi.org/10.1517/14740338.2014.913021
32. Ferrari SM, Fallahi P, Galetta F et al (2018) Thyroid disorders induced by checkpoint inhibitors. Rev Endocr Metab Disord 19(4):325–333. https://doi.org/10.1007/s11154-018-9463-2
33. Antonelli A, Wartofsky L, Miccoli P (2018) Editorial: levothyroxine therapy in patients with hypothyroidism. Front Endocrinol (Lausanne) 12:734895. https://doi.org/10.3389/fendo.2021.734895
34. Aoki Y, Belin RM, Clickner R et al (2007) Serum TSH and total T4 in the United States population and their association with participant characteristics: National Health and nutrition examination survey (NHANES 1999-2002). Thyroid 17(12):1211–1223. https://doi.org/10.1089/thy.2006.0235

35. Persani L (2012) Clinical review: central hypothyroidism: pathogenic, diagnostic, and therapeutic challenges. J Clin Endocrinol Metab 97(9):3068–3078. https://doi.org/10.1210/jc.2012-1616
36. Chaker L, Razvi S, Bensenor IM et al (2022) Hypothyroidism. Nat Rev Dis Primers 8:30. https://doi.org/10.1038/s41572-022-00357-7
37. Wartofsky L (2006) Myxedema coma. Endocrinol Metab Clin N Am 35(4):687–98, vii–viii. https://doi.org/10.1016/j.ecl.2006.09.003
38. Nygaard B (2010) Hypothyroidism (primary). BMJ Clin Evid 2010:0605
39. Kravets I (2016) Hyperthyroidism: diagnosis and treatment. Am Fam Physician 93(5):363–370
40. Visser WE, van Mullem AA, Visser TJ (2013) Different causes of reduced sensitivity to thyroid hormone: diagnosis and clinical management. Clin Endocrinol 79(5):595–605. https://doi.org/10.1111/cen.12281
41. Schoenmakers E, Agostini M, Mitchell C et al (2010) Mutations in the selenocysteine insertion sequence-binding protein 2 gene lead to a multisystem selenoprotein deficiency disorder in humans. J Clin Invest 120(12):4220–4235. https://doi.org/10.1172/JCI43653
42. Bernal J, Guadaño-Ferraz A, Morte B (2015). Thyroid hormone transporters–functions and clinical implications. Nat Rev Endocrinol 11(7):406–417. doi:https://doi.org/10.1038/nrendo.2015.66
43. Guillén-Yunta M, Valcárcel-Hernández V, García-Aldea Á et al (2015) Neurovascular unit disruption and blood-brain barrier leakage in MCT8 deficiency. Fluids Barriers CNS. 3 Nov 2023 20(1):79. https://doi.org/10.1186/s12987-023-00481-w

Chapter 2

Thyroid Hormone Action by Genomic and Nongenomic Molecular Mechanisms

Ana Aranda

Abstract

The thyroid hormones, thyroxine (T4) and triiodothyronine (T3), are pivotal in regulating various physiological processes including growth, development, and metabolism. The biological actions of thyroid hormones are primarily initiated by binding to nuclear thyroid hormone receptors (TRs). These receptors, belonging to the superfamily of nuclear receptors, act as ligand-dependent transcription factors. Transcriptional regulation by TRs is mediated by the recruitment of coregulators, governing activation and repression of target genes, thereby modulating cellular responses to thyroid hormones. Beyond this canonical genomic pathway, TH can regulate the expression of genes not directly bound by TRs through cross-talk mechanisms with other transcription factors and signaling pathways. Thyroid hormones can also elicit rapid non-genomic effects, potentially mediated by extranuclear TR proteins or by interactions with membrane receptors such as integrin αvβ3. This non-genomic mode of action adds another layer of complexity to the diverse array of physiological responses orchestrated by thyroid hormones, expanding our understanding of their multifaceted actions.

Key words Thyroid hormones (THs), Thyroid hormone receptors (TRs), Integrin αvβ3 membrane receptor, Thyroid hormone response elements

1 Nuclear Thyroid Hormone Receptors (TRs)

The thyroid hormones, thyroxine (T4) and triiodothyronine (T3), are pivotal in numerous physiological processes, encompassing growth, development and metabolism, thereby exerting profound effects on the body's homeostasis, and are also implicated in various pathological processes. The majority of the biological actions of the thyroid hormones (THs) are mediated by binding to the nuclear thyroid hormone receptors (TRs). TRs belong to the superfamily of nuclear receptors that function as ligand-dependent transcription factors [1]. In addition to TRs, the nuclear receptor superfamily comprises receptors for other small lipophilic molecules such as steroid hormones, vitamin A, vitamin D, cholesterol derivatives, and other metabolites, as well as "orphan" receptors with unknown

Constanza Contreras-Jurado (ed.), *Thyroid Hormones: Methods and Protocols*, Methods in Molecular Biology, vol. 2876, https://doi.org/10.1007/978-1-0716-4252-8_2,

ligands. Although the thyroid gland predominantly produces T4, T3 is the most active hormone due to its much higher affinity for the nuclear receptors [2].

TRs are encoded by two genes, *TRα* and *TRβ*, which in humans are located in chromosomes 17 and 3, respectively. The *TRα* and *TRβ* genes generate several TR protein isoforms through promoter usage or alternative splicing of primary transcripts. The TRα1 and TRα2 isoforms are produced by alternative splicing of the *TRα* gene and differ at the carboxy-terminus. TRα1 is a genuine nuclear receptor, while TRα2 cannot bind hormone with high affinity and therefore does not function as a receptor. Instead, it may have an antagonistic effect on TR actions [3]. Other minor four proteins (p43, p33, p30, and p28) have been identified, which are encoded in the same mature mRNA that also encodes TRα1. The truncated p30 receptor was found to be inserted into the plasma membrane [4], while p43 and p28 can be imported into the mitochondria [5]. By usage of different transcription start points the *TRβ* gene gives rise to two main hormone-binding isoforms β1 and β2, with different amino terminus. Other minor TRβ isoform TRβ3, as well as truncated forms with dominant negative activity, which cannot bind DNA, have been also identified in rodents although their functions have not yet been defined [2, 6].

The expression patterns of the main hormone-binding isoforms, TRα1, TRβ1, and TRβ2, differ across various cell types and developmental periods, suggesting their potential role in cell and tissue-specific functions [2]. The distribution of these TR isoforms can significantly influence the specific responses to THs in diverse tissues and organs. TRα1 is predominantly found in the heart, brain, and muscle tissues, among others, where it contributes significantly to postnatal growth and cardiac functionality [7]. In contrast, TRβ1 exhibits a wider expression, mostly in the kidney and liver, where it governs metabolic processes. It is also highly expressed in the pituitary, regulating the levels of circulating thyroid hormones via the hypothalamic-pituitary-thyroid axis. On the other hand, the presence of TRβ2 is restricted to the anterior pituitary and specific neural cells, having a notable influence on the development of the retina and the inner ear [8].

TRs and other nuclear receptors have a modular structure with distinct regions corresponding to autonomous functional domains. TRs typically include a variable N-terminal region (A/B), a conserved DNA-binding domain (region C or DBD), a linker region (D), and a conserved region (E) containing the ligand binding domain (LBD) [1]. The A/B region is the most variable both in terms of its size and sequence and contains a ligand-independent transcriptional activation function (AF1). On the other hand, the DBD confers the ability to recognize and bind with high affinity to the thyroid hormone response elements or TREs. This domain consists of two "zinc fingers" and a carboxy-terminal extension.

In every zinc finger, four invariable cysteines tetrahedrally coordinate a single zinc ion. The amino acids crucial for discrimination of the TRE are located at the base of the first finger in an area designated as the "P box". Additionally, other residues of the second zinc finger, which form the so-called "D box", participate in dimerization. Structural studies have revealed that the core DBD is made up of two α helices. The first one, known as the recognition helix, binds to the DNA's major groove and includes the "P box" that interacts with specific DNA bases of the TRE. The second helix encompasses the "D box" and forms a right angle with the recognition helix [9].

The D domain in TRs functions as a hinge between the DBD and the LBD, enabling rotation of the DBD. The D domain contains residues whose mutation eliminates interaction with nuclear receptor corepressors. Furthermore, this region strongly associates with the LBD, but only in the presence of ligand or corepressors, providing stabilization of the receptor's overall structure [10].

In addition to hormone binding, the LBD serves multiple functions, including mediating homo and heterodimerization, as well as ligand-dependent transcriptional activation and repression. The LBDs contain a carboxy-terminal AF2 motif that is responsible for ligand-dependent transcriptional activation. The crystal structure of the LBD reveals that it is composed of twelve conserved α-helical regions, numbered from H1 to H12. The hormone-binding pocket is primarily composed of non-polar amino acids and is located in the lower half of the LBD. The AF2 domain, located in H12, is necessary for ligand-dependent transactivation. Although the core AF2 activity is present in H12 of the LBD, this domain also includes other dispersed elements that come together upon hormone binding. One such element is a region that includes the C-terminal half of helix 3 and helix 4. Mutations in this region do not affect ligand binding or dimerization but do impair ligand-dependent transactivation. Specifically, a highly conserved lysine in the carboxy-terminus of H3 is crucial for transcriptional activity not only in TR but also in most nuclear receptors [1].

The ligand-bound TRs are more compact than the unliganded ones, demonstrating that the receptors undergo a clear conformational change upon hormone binding. The most striking difference observed upon hormone binding is the position of H12 [11]. The helix extends outward from the LBD when no ligand is present. However, when the hormone binds, H12 shifts in a "mouse-trap" model, tightly packing against H3 or H4 and directly contacting the hormone. This alteration creates a hydrophobic cleft that interacts with coactivators [12].

2 Transcriptional Regulation by TRs

The transcriptional regulation orchestrated by TRs involves their association with the retinoid X receptor (RXR) and binding to thyroid hormone response elements (TREs) within the regulatory regions of target genes. Although initially it was believed that TRs bind to TREs located in target gene promoters, mapping of the genome-wide binding sites by ChIP-seq has shown that the receptor binds to a large number of genomic sites far from known transcription start sites. Transcriptional regulation by TRs is mediated by the recruitment of coregulators, coactivators, and corepressors, governing activation and repression of genes in a hormone-dependent manner, thereby modulating cellular responses to thyroid hormones.

2.1 Binding to TREs and Dimerization

TRs bind preferentially as heterodimers with other nuclear receptor, the retinoid X receptor (RXR), to TREs composed of two copies of the AGG/TTCA motif. These motifs are configured as palindromes (Pal), inverted palindromes (IPs) or, most commonly, as direct repeats separated by 4 non-conserved base pairs (DR4) [2]. Although TRs can bind to their response elements as homodimers, heterodimerization with RXR significantly increases the affinity for DNA and transcriptional activity. DRs are inherently asymmetric, so heterodimeric complexes may bind to them in two distinct polarities. However, it has been established that on a DR4, RXR occupies the upstream half-site and TR is situated in the downstream motif [13]. The TR/RXR heterodimer's capacity to bind to palindromes, IPs, and DR elements suggests that the DBDs must be rotationally flexible in relation to the LBDs dimerization interface. This is mediated by the hinge domain. Additionally, a distinct region of the DBD of each receptor must be employed to establish the dimerization interface in the different TREs. It was believed that the formation of the TR/RXR heterodimer prevented the binding of ligand to RXR. In these complexes, RXR was thought to be a "silent partner" with the sole function of increasing binding to the TRE. However, RXR can bind ligand and recruit coactivators as a heterodimer with TR. This finding suggests that in certain cellular environments, RXR can act as a "non-silent" partner of TR, enabling stimulation by RXR agonists [14].

2.2 Thyroid Hormone-Dependent Transcriptional Activation: Coactivators

Ligand-dependent transcriptional stimulation by nuclear receptors is mediated by the recruitment of coactivators. As stated above, repositioning of the ligand-dependent transcriptional activation function (AF2) plays a fundamental role in coactivators binding. Some of these proteins are chromatin-remodeling factors or possess histone-modifying activity, whereas others may interact directly with the basic transcriptional machinery. Cyclical recruitment of all these coactivator complexes to the target gene causes chromatin remodeling, the association of the basal transcriptional machinery, and transcriptional activation. Although the knowledge about the mechanisms of transcriptional activation by nuclear receptors and the role of the different coregulators has greatly increased in the last years, there are still numerous gaps about the role of different proteins on gene regulation by nuclear receptors, mainly in models of natural complex promoters or pathophysiological processes.

Expression of genes in higher organisms is critically dependent on DNA accessibility. Two major mechanisms alleviate the block of transcription caused by the nucleosomal structure: histones can be post-translationally modified to destabilize chromatin, and nucleosomes can be disrupted through the activity of ATP-dependent chromatin remodeling complexes [15]. Posttranslational modifications of the carboxy-terminal histone "tail" domains play an intrinsic role in transcriptional activation and/or repression, controlling the folding of nucleosomal arrays into higher-order structures. These covalent alterations are believed to form a "histone code" in which specific combinatorial sets of histone modification signals can dictate transcriptional activation or repression [16]. Chromatin of transcriptionally active genes has been known for a long time to be enriched in hyperacetylated histones. Histone acetylation is catalyzed by histone acetyltransferases (HATs), whereas acetyl groups are removed from the histone tails by the action of histone deacetylases (HADCs). Histone acetylation is a critical step in TR-mediated hormone signaling. The receptors do not have HAT enzymatic activity. However, coactivator complexes of the SRC/p160 family with HAT activity are recruited to TRs in a ligand-dependent manner [17]. The p160 coactivators have a nuclear receptor-interacting domain (RID) in their central region with three highly conserved LxxLL motifs, where L is leucine and x is any amino acid. This domain mediates the association of the coactivators to hormone-bound receptors. The cocrystal structure of the LBD of TR and a p160 coactivator fragment containing two LxxLL motifs shows that a conserved glutamic acid in H12 and the lysine in H3 of the receptor LBD form hydrogen bonds with leucines 1 and 5 of the coactivator RID. These contacts create a charge clamp that orients and positions the coactivator RID into the hydrophobic groove formed in the LBD after the conformational change caused by ligand binding [12, 18]. This is why mutations of these residues in TRs result in a transcriptionally inactive receptor. The in vivo importance of the coactivators and the AF2 domain is demonstrated by the finding that mice homozygous for the knock-in mutation at glutamic acid residue in H12 of TRβ, resulting in disruption with p160 coactivators, have high serum TSH and TH levels, consistent with resistance to TH. This also demonstrates that the AF2 domain is required for both up-regulated and down-regulated TH action [19].

In addition to possessing HAT activity, p160 coactivators also function as platforms for recruiting secondary coactivators to the TRs. p160 coactivators contain two domains with intrinsic transcriptional activity. The stronger transactivation domain (AD1) is the region that interacts with the HAT CBP/p300, which serves as a coactivator for nuclear receptors and many other transcription factors, acting as a cointegrator of extracellular and intracellular signals. Additionally, p160 coactivators associate with P/CAF, the first identified mammalian HAT. P/CAF interacts directly and independently with receptors and CBP/p300 through different regions and acts as a nuclear receptor coactivator. Therefore, p160 coactivators may serve as a docking platform to bridge protein complexes with HAT activity to DNA-bound TRs. Both P/CAF and CBP interact with TBP-associated factors (TAFs) and RNA polymerase II. Thus, they have dual functional roles as adaptors of the receptors to the basal transcriptional machinery and as enzymes that can modify chromatin structure through HAT activity [1]. It has been demonstrated that in vivo histone acetylation of TH target genes is strongly induced upon treatment with the hormone [20]. However, histone hyperacetylation induced by hormones at target genes is transient and cyclic. The proposed mechanism is that p160 coactivators can be acetylated by CBP/p300. Acetylation neutralizes the positive charges of two lysine residues adjacent to the core LxxLL motif and disrupts the association of HAT coactivator complexes with the receptors [21].

Protein arginine methyltransferases (PRMTs) have been linked to gene activation through histone tails methylation [22]. PRMT1 and CARM1, also known as PRMT4, act as secondary coactivators in nuclear hormone receptor-regulated gene expression by binding to the AD2 region of the p160 coactivators. Their coactivator potential depends on an intact histone methyltransferase (HMT) activity [23]. CARM1, p160, and CBP/p300 work together to increase transcriptional activation by nuclear receptors in a ligand-dependent manner [24].

Nuclear receptors can bind to their response elements packaged in chromatin, enabling the association of other factors whose binding sites were not previously exposed in the nucleosomes. This is determined by ATP-dependent remodeling complexes, such as SWI2/SNF2 or ISWI/SNFL2. These complexes use the energy from ATP hydrolysis to catalyze nucleosome mobilization, which results in a net change in the position of the histone octamer relative to the DNA. It is thought that this change facilitates the access and function of key components of the transcriptional machinery. Chromatin remodeling complexes consist of an ATPase subunit (BRG-1 or hBrahma) together with BRG-1-associated factors or BAFs [25]. The receptors do not interact directly with BRG-1 or hBrahma, but BAF57 binds the N-terminal domain of p160 coactivators and different receptors interact with different

BAF subunits. Thus, a direct interaction between BAF57, a key SWI/SNF subunit, and TRβ has been observed at T3-responsive promoters [26]. In this way, hormone binding causes the recruitment of these complexes to chromatin, leading to nucleosome remodeling. Indeed, it has been recently shown that selective recruitment of SWI/SNF complexes to TRβ binding sites regulates chromatin accessibility and gene expression [27].

The transcription of RNA polymerase II-dependent genes requires binding of the general transcription factors (GTFs) to the transcription initiation site. A complex consisting of the polymerase, GTFs, SRBs (suppressors of RNA polymerase B), and several other proteins, termed "the holoenzyme", can be directly recruited to gene promoters by sequence-specific transcription factors to initiate transcription. The Mediator complex, which associates to the large subunit of RNA polymerase II, is responsible for the recruitment of the "holoenzyme" to the target promoter. The Mediator complex is recruited to the AF2 TR core region in response to hormone binding by a single subunit (MED1 or TRAP220), which contains an LxxLL motif identical to that found in the p160 coactivators, thereby enhancing the hormone-dependent transcriptional activity [28]. Moreover, MED1 also promotes nuclear retention of TR, showing the dual functionality of H12, which serves to modulate both gene transactivation and nuclear TR anchoring [29].

TH-dependent transcriptional activity requires recruitment of both the Mediator and p160 coactivator complexes to TR. Since both interact with the same receptor region they may compete with each other for binding to the receptor. Studies using chromatin immunoprecipitation (ChIP) assays have shown that p160 and Mediator complexes are recruited to hormone-regulated promoters in a sequential manner [30]. p160 complexes act earlier than the Mediator and this ordered recruitment can continue for multiple cycles of factor association and dissociation, with multiple rounds of transcription occurring within each cycle [31].

In summary, the binding of hormone to TRs allows for the recruitment of coactivators with unique biochemical activities at temporally appropriate times during the transcriptional process. The receptors can first remodel chromatin and recruit coactivators with histone-modifying activity, resulting in chromatin decondensation. Once chromatin is no longer condensed, the receptors through their association with the Mediator complex would be able to recruit the transcriptional machinery leading to stimulation of target gene expression.

2.3 Transcriptional Repression by Unliganded Receptors: Corepressors

In the absence of hormone TRs can function as transcriptional repressors instead of acting as inactive, passive receptors because they bind to corepressors such as NCoR1 (nuclear receptor corepressor (1) or SMRT (silencing mediator of retinoic and thyroid receptor or NCoR2). Unliganded TRs strongly interact with NCoR1 and SMRT and addition of hormone induces dissociation from the corepressors. NCoR1 and SMRT are both structurally and functionally related. They contain three autonomous repressor domains (RD) and a receptor-interacting domain (RID) located toward the carboxyl terminus. The RID consists of two motifs (or CoRNR boxes), with the consensus sequence Lxx I/H I xxx I/L [32]. Although a receptor CoR box, located in H1 of the LBD within the TR hinge region, is essential for corepressors binding, the CoRNR box does not interact directly with residues in this region, but instead docks to a hydrophobic groove in the surface of the LBD helices 3 and 4. Because this surface overlaps with that involved in coactivator interaction, coactivator and corepressor binding are mutually exclusive. Ligand binding alone is not sufficient to induce corepressor dissociation. Rather, it appears that the AF2 region is involved in triggering the release of corepressors from the receptors. Indeed, deletion of the TR AF2 domain increases corepressor interaction and also reduces corepressors release following hormone binding, indicating that H12, opposite to that observed with coactivators, is inhibitory for corepressor binding [33].

Histone deacetylation plays a crucial role in repression by unliganded TR, since corepressors belong to multicomponent repressor complexes that contain histone deacetylases (HDACs) and cause chromatin condensation. Corepressors recruit HDAC1 or 2 (class I deacetylases) to the carboxyterminal repressor domain RD1 via the adaptor mSin3 protein, and directly interact through the RD3 domain with class II deacetylases (HDACs 4, 5, and 7). Therefore, a single corepressor could use distinct domains to engage class I HDAC complexes in a Sin3-dependent manner and class II HDAC complexes in a Sin3-independent manner [34, 35]. Furthermore, a repressor complex containing the corepressors, HDAC3 and transducin beta-like proteins (TBL1 or TBL1R), appears to be required for repression by TR [36, 37]. Knock-in mice with mutations in the receptor interaction domain (RID) of SMRT and NCoR that abolish interaction with TRs show their importance in TR signaling [38]. Surprisingly, transcriptional repression of a number of target genes by the unliganded TR can occur in the absence of NCoR1, but is still associated with decreased acetylation in the region of TR-binding sites. This indicates that TRs could recruit still unidentified factors additionally to NCoR1 or SMRT to mediate its ligand-independent effects [39].

2.4 Models of Thyroid Hormone-Dependent Transcriptional Activation

The classical model for hormone-dependent activation by TRs is a bimodal "all or none" switch in which the unliganded TR would be bound to chromatin and by interaction with corepressors complexes containing HDACs silences transcription of target genes. Binding of hormone would cause corepressors release and facilitate the recruitment of the different types of coactivator complexes responsible for transcriptional activation [32]. However, it has been reported that, while the bimodal switch applies for some genes, a considerable number of genome-wide TR binding sites may be formed de novo in the liver in the presence of hormone [40]. This supports a second model in which TR is not bound to chromatin in the absence of TH, but is recruited to chromatin in a ligand-dependent manner, inducing de novo chromatin remodeling. It has been proposed that the presence of a strong TRE, a DR4 motif, allows TR occupancy of accessible chromatin by the unliganded receptor. However, chromatin harboring less stringent TR-binding motifs would not be accessible to the unliganded TRs and would be only bound by the receptors in the presence of hormone. This ligand-facilitated recruitment may be orchestrated by cooperation with other bound transcription factors and remodeling proteins. In contrast, a more recent genome-wide study showed robust TR binding to DR4 motifs in chromatin in the absence of hormone that was further stimulated by TH, rather than de novo formation of TR binding sites [41]. Similarly, it has been recently shown that in the pituitary T3 regulates histone modifications and chromatin opening primarily at sites in which TR binds independently of T3 levels rather than at sites where T3 induces de novo binding [42]. Further studies focused on enhancers' hyperacetylation in response to hormone have recently shown that some enhancers containing TREs are hyperacetylated by release of HDACs, while others are hyperacetylated in response to T3-dependent TR and coactivators recruitment. Interestingly, a subset of enhancers hyperacetylated in response to hormone do not contain TR-binding sites. In this case, these enhancers may come to close proximity to TR-bound enhancers by long-range interactions within super-enhancers or topologically associated domains [20]. Moreover, recently a shift in the coactivators to corepressors ratio bound to chromatin, rather than the canonical "all or none" switch has been described, since NCoR1/HDAC3 corepressor complexes were decreased but not completely dismissed by TH at TR-bound liver chromatin regions [41]. Collectively, this demonstrates that various mechanisms control transcriptional responses to TH, which would be more complex than the simple bimodal switch model.

In addition to TH-dependent regulation of gene expression mediated by binding of TRs to TREs, non-canonical TRE-independent signaling by TRs has been demonstrated in knock-in mice in which wild-type receptors have been replaced by mutants in the

P-box of the DBD. This mutation abrogates binding to TREs, leading to loss of canonical TH actions. In contrast to TRβ knockout mice, these mutant mice maintain normal hearing at certain frequencies and do not display significant outer hair cell loss [43]. However, DNA-binding is essential for many functions of the TR, including retinal development and negative feedback regulation by TH of the hypothalamic-pituitary-thyroid axis. Moreover, several effects of THs linked to energy metabolism, body temperature, locomotor activity, and heart rate are still preserved in these mice, providing in vivo evidence that TRs mediate physiologically relevant effects that are independent of binding to TREs [44]. Whether these mutations, in addition to prevent binding to the TRE, affect other TR functions, for instance altering protein to protein interactions, remains to be studied.

2.5 Thyroid Hormone-Dependent Transcriptional Repression

In addition to mediating TH-dependent transcriptional activation, TRs can also repress gene transcription in a hormone-dependent manner. Indeed, expression of many genes is inhibited by T3, among them the TSH and TRH genes, indicating the importance of transcriptional repression in the feed-back mechanism of the hypothalamus-pituitary-thyroid axis.

In some cases, hormone-dependent repressive effects may be due to passive inhibition, which can occur due to competition for DNA binding motifs with other transactivators, to formation of transcriptionally inactive heterodimers or to competition of liganded TRs for common transcriptional cofactors (squelching) that can result in their release from DNA and to reduced gene transcription [1]. However, in other cases this repression might be associated to binding to negative TREs. Although the properties of these motifs are not totally understood, location of the element may play a role since they are generally located very close to the transcription initiation site, either upstream or downstream, or even have an unusual location at the 3′-untranslated region. However, TR binding to putative TREs may be weak in comparison to the binding sites in positively regulated genes. Indeed, whole genome analysis of TR binding sites in chromatin has shown that, in contrast to positively regulated target genes, many T3-negative targets do not bind TR directly [40], implying that they might be regulated by indirect mechanisms. Nevertheless, it has also been reported that two TRE half-sites without spacing (DR0) can be found close to some of the T3-repressed genes, implying a direct TR binding [45], and a more recent study has shown that TH-dependent gene repression correlates with direct binding of TR to DR4-like motifs on a genome-wide scale. In contrast with positively regulated genes, chromatin accessibility decreases upon TH treatment at these motifs near down-regulated genes [41].

Corepressors and deacetylase activity appear to be involved in hormone-dependent negative regulation in vivo. Indeed, a knock-in mouse harboring the R429Q mutation that abolishes corepressor binding shows selective impairment of TH-mediated gene repression while preserving gene stimulation, suggesting that corepressors have a critical role in negative gene regulation by TH [46]. Furthermore, a NCoR1 knock-in mouse in which binding to HDAC3 is genetically disrupted shows that the interaction of the corepressor with the deacetylase is also important for the control of negatively-regulated genes by TH in vivo [47]. Coactivators appear to play an apparently paradoxical role in T3-dependent negative regulation of gene expression. A knock-in mouse expressing a TRβ mutant in H12 also shows defective negative regulation of TSH, indicating that gene repression by TH requires an intact coactivator-binding surface [19]. Furthermore, deletion of SRC-1 results in marked increases of serum TH and TSH levels, showing a phenotype of resistance to TH [48], while SRC-1 knockout mice that also express a mutant NCoR1 that cannot interact with TRβ display normal TH and TSH levels, indicating that a critical balance between coactivators and corepressors is critical in gene repression by TH in vivo [49]. These mutational studies agree with the recent report showing that TH reduces (but do not abolish) binding of the coactivator CBP at TR binding motifs near down-regulated genes, and that the CBP to NCoR1 (or HDAC3) ratio is higher at down-regulated TR-bound enhancers in the absence of hormone. This again reinforces the idea that in contrast to the canonical "all or none" coregulator switch model, TH orchestrates a shift in the relative binding of corepressors and coactivators to not only positively but also negatively regulated target genes [41].

TRs can also inhibit expression of genes that do not contain a TRE by negative regulation of the activity of other transcription factors or signaling pathways, a mechanism normally referred to as transcriptional crosstalk. Thus, the receptors can negatively regulate target gene promoters that carry binding sites for the transcription factors AP1 [50], NF-κB [51], CREB [52], SMAD [53], STAT3 [54], or GATA-2 [55] without binding to these DNA elements themselves. The mechanisms responsible can involve direct protein to protein interaction of the receptors with these transcription factors, or with intermediary factors that tether the receptors to the regulated promoter. The receptors can also regulate the activity of the signaling pathways that lead to activation of these factors by different mechanisms. It is believed that many of the antiproliferative effects and anti-inflammatory actions of ligands of different nuclear receptors could be mediated by transcriptional "cross-talk" mechanisms.

3 Non-genomic Extranuclear TR Actions

Emerging evidence suggests that THs can also exert extranuclear non-genomic effects occurring after seconds to minutes, thus precluding transcription and protein synthesis, through a variety of mechanisms still not yet fully clarified. These effects may be related to the localization of a fraction of TRs outside the nucleus, often involving activation of phosphorylation cascades. It should be, however, noted that activation of these cascades can also lead to changes in gene transcription. The more convincing non-genomic TR effects are mediated by phosphoinositol 3-kinase (PI3K) activation. The PI3K/Akt (protein kinase B) pathway is an important regulator of different cellular functions. TH stimulation of the PI3K pathway happens within minutes, cannot occur in the absence of TRs, and appears to involve a direct interaction of either TRα or TRβ with the p85 regulatory subunit of PI3K [56–59]. Induction in a TRβ-specific manner of hypoxia-inducible factor 1α (HIF-1 α), and its glucose metabolism target genes in fibroblasts has been described to be mediated by PI3K activation [60]. TRβ is able to interact not only with the p85 subunit but also with Lyn, a PI3K activator of the Src family of membrane tyrosine kinases. In the presence of TH, the receptor dissociates from this complex and PI3K is activated. Interaction with Lyn depends on a canonical IYVGM motif present in the second zinc finger of TRβ. Mutation of tyrosine 147 to phenylalanine (Y147F) abolishes rapid TH signaling through PI3K but does not affect TH-dependent transcription of TRE-containing genes. Importantly, TRβ Y147F knock-in mice show impaired maturation and plasticity of hippocampal synapsis, indicating that noncanonical PI3K signaling pathway activation is involved in this process [61]. However, these mice display a normal hepatic gene regulation and proliferative response indicative of a canonical TRβ-dependent mechanism [62].

Non-genomic TR effects have been also attributed to extranuclear TR variants. The 30-kD TRα isoform generated from alternative translation was shown to be palmitoylated and associated with caveolin-containing plasma membrane domains [4]. Although this protein lacks the DBD and is transcriptionally inactive, TH binding to this N-terminal truncated receptor leads to increased intracellular calcium, nitrous oxide, and cyclic guanine monophosphate (cGMP), which in turn activates the cGMP protein kinase II (PKGII), Src, extracellular-signal-regulated kinase (ERK), and PI3K/Akt signaling pathways. PI3K/AKT stimulation is involved in the acute vasodilatory and neuroprotective effects of TH [58], as well as in osteoblasts proliferation and survival [4]. Despite stimulation of these pathways, p30 TRα is necessary, but not sufficient, for T3-induced osteoblast proliferation, suggesting cooperation between membrane-initiated and nuclear T3 signaling in bone physiology [4].

It has been known by a very long time that TH has profound effects on mitochondrial activity and cellular energy state. In addition to genomic actions, THs may mediate direct effects on mitochondria by binding to TRs located in this organelle. While full-length TRα1 contains at least two nuclear localization signals (NLS) as well as atypical mitochondrial import signals [63], the N-terminally truncated forms of TRα p43 and p28 lack one or both NLS and are translocated to mitochondria [64]. While p28 TRα is associated with the inner mitochondrial membrane, TRα p43 is localized to the matrix [65]. The function of p28 TRα is presently unknown, but p43 TRα can bind DNA with an affinity similar to that of the full-length receptor and increase mitochondrial activity and mitochondriogenesis [66]. The mitochondrial genome contains four TRE-like sequences and these effects can be abolished by deletion of the DBD, suggesting that p43 TRα may act as a T3-dependent transcription factor of the mitochondrial genome [67].

4 Non-genomic TR-Independent Thyroid Hormone Actions

In addition to the mechanisms of action already discussed, some non-genomic TH functions might be also TR-independent. The existence of TH receptors at the plasma membrane, different from TRs, had been proposed and a hormone binding site in integrin αvβ3 has been identified [68]. Integrins are a family of transmembrane receptors that mediate cell to cell and cell to intracellular matrix interaction, thus having important cellular functions. Integrin αvβ3 is mainly expressed in endothelial cells, osteocytes, tumor cells, and some hematopoietic cells, with little expression in other cell types. In contrast with TRs that bind preferentially T3, integrin αvβ3 shows a higher affinity for T4. The integrin can also bind the TH metabolite tetraiodothyroacetic acid (TETRAC), naturally derived from T4, which can compete with T4 binding acting as an antagonist [69, 70]. Computational modeling of T4 and T3 binding to the integrin is consistent with the existence of two TH binding domains, named S1 and S2 [71]. S1 binds exclusively T3, whereas S2 can bind both T4 and T3, with a lower affinity for T3 [72]. Binding of T4 to domain S2 activates the ERK signal transduction pathway, which could promote cell proliferation, inhibit apoptosis, and stimulate angiogenesis. However, T3 interacts with the S1 domain to activate the Src/PI3K/AKT pathway, which does not appear to regulate cell proliferation, but can alter the shuttling from the cytoplasm to the nucleus of TRα and other proteins [73].

The possibility that the alternate T3 signaling through integrin αvβ3 requires the presence of full-size or truncated TR isoforms cannot be dismissed. Thus, T3-induced activation of ERK and PI3K/Akt in osteoblasts by binding to p30 TRα occurs through activation of PKGII and requires the integrin β3 subunit [4]. PKGII activates Src bound to the cytoplasmic tail of β3 integrin subunits, leading to ERK and Akt activation [74]. This may explain why antibodies against integrin αvβ3 subunits and siRNA-mediated depletion of β3 can block TH-induced ERK and Akt activation [68, 72]. On the other hand, TH-dependent regulation of brain actin polymerization in the developing brain [75], as well as induction of migration and neurite outgrowth in cultured brain cells [76] has been described to rely on a non-genomic mechanism. However, it has been recently described that although TH signaling via integrin αvβ3 is required to induce ERK and PI3K phosphorylation, as well as dendrite arborization, this effect was suppressed not only by knockdown of the integrin, but also of TRα and TRβ [77]. Therefore, more cellular models and genetically modified animal models are needed to sort out the genomic and non-genomic actions of the TH.

In conclusion unraveling the intricate mechanisms of transcriptional regulation by TRs, as well as the interplay of the genomic and non-genomic actions of the thyroid hormones, will be essential for understanding their overall physiological effects and for the development of novel treatments for various pathological conditions controlled by these hormones.

References

1. Aranda A, Pascual A (2001) Nuclear hormone receptors and gene expression. Physiol Rev 81(3):1269–1304
2. Aranda A, Alonso-Merino E, Zambrano A (2013) Receptors of thyroid hormones. Pediatr Endocrinol Rev 11(1):2–13
3. Hones GS, Harting N, Mittag J, Kaiser FJ (2022) TRalpha2-an untuned second fiddle or fine-tuning thyroid hormone action? Int J Mol Sci 23(13). https://doi.org/10.3390/ijms23136998
4. Kalyanaraman H, Schwappacher R, Joshua J, Zhuang S, Scott BT, Klos M et al (2014) Non-genomic thyroid hormone signaling occurs through a plasma membrane-localized receptor. Sci Signal 7(326):ra48. https://doi.org/10.1126/scisignal.2004911
5. Blanchet E, Pessemesse L, Feillet-Coudray C, Coudray C, Cabello C, Bertrand-Gaday C et al (2021) A truncated form of thyroid hormone receptor alpha, regulates maturation of pancreatic beta cells. Int J Mol Sci 22(5):43. https://doi.org/10.3390/ijms22052489
6. Refetoff S, Dumitrescu AM (2007) Syndromes of reduced sensitivity to thyroid hormone: genetic defects in hormone receptors, cell transporters and deiodination. Best Pract Res Clin Endocrinol Metab 21(2):277–305. https://doi.org/10.1016/j.beem.2007.03.005
7. Fraichard A, Chassande O, Plateroti M, Roux JP, Trouillas J, Dehay C et al (1997) The T3R alpha gene encoding a thyroid hormone receptor is essential for post-natal development and thyroid hormone production. EMBO J 16(14):4412–4420. https://doi.org/10.1093/emboj/16.14.4412
8. Forrest D, Vennstrom B (2000) Functions of thyroid hormone receptors in mice. Thyroid 10(1):41–52. https://doi.org/10.1089/thy.2000.10.41
9. Rastinejad F, Perlmann T, Evans RM, Sigler PB (1995) Structural determinants of nuclear receptor assembly on DNA direct repeats. Nature 375(6528):203–211. https://doi.org/10.1038/375203a0
10. Pissios P, Tzameli I, Kushner P, Moore DD (2000) Dynamic stabilization of nuclear receptor ligand binding domains by hormone or corepressor binding. Mol Cell 6(2):245–253.

https://doi.org/10.1016/s1097-2765(00)00026-5
11. Wagner RL, Apriletti JW, McGrath ME, West BL, Baxter JD, Fletterick RJ (1995) A structural role for hormone in the thyroid hormone receptor. Nature 378(6558):690–697. https://doi.org/10.1038/378690a0
12. Feng W, Ribeiro RC, Wagner RL, Nguyen H, Apriletti JW, Fletterick RJ et al (1998) Hormone-dependent coactivator binding to a hydrophobic cleft on nuclear receptors. Science 280(5370):1747–1749. https://doi.org/10.1126/science.280.5370.1747
13. Kurokawa R, Yu VC, Naar A, Kyakumoto S, Han Z, Silverman S et al (1993) Differential orientations of the DNA-binding domain and carboxy-terminal dimerization interface regulate binding site selection by nuclear receptor heterodimers. Genes Dev 7(7B):1423–1435. https://doi.org/10.1101/gad.7.7b.1423
14. Castillo AI, Sanchez-Martinez R, Moreno JL, Martinez-Iglesias OA, Palacios D, Aranda A (2004) A permissive retinoid X receptor/thyroid hormone receptor heterodimer allows stimulation of prolactin gene transcription by thyroid hormone and 9-cis-retinoic acid. Mol Cell Biol 24(2):502–513. https://doi.org/10.1128/MCB.24.2.502-513.2004
15. Narlikar GJ, Fan HY, Kingston RE (2002) Cooperation between complexes that regulate chromatin structure and transcription. Cell 108(4):475–487. https://doi.org/10.1016/s0092-8674(02)00654-2
16. Jenuwein T, Allis CD (2001) Translating the histone code. Science 293(5532):1074–1080. https://doi.org/10.1126/science.1063127
17. Glass CK, Rosenfeld MG (2000) The coregulator exchange in transcriptional functions of nuclear receptors. Genes Dev 14(2):121–141
18. Darimont BD, Wagner RL, Apriletti JW, Stallcup MR, Kushner PJ, Baxter JD et al (1998) Structure and specificity of nuclear receptor-coactivator interactions. Genes Dev 12(21):3343–3356. https://doi.org/10.1101/gad.12.21.3343
19. Ortiga-Carvalho TM, Shibusawa N, Nikrodhanond A, Oliveira KJ, Machado DS, Liao XH et al (2005) Negative regulation by thyroid hormone receptor requires an intact coactivator-binding surface. J Clin Invest 115(9):2517–2523. https://doi.org/10.1172/JCI24109
20. Praestholm SM, Siersbaek MS, Nielsen R, Zhu X, Hollenberg AN, Cheng SY et al (2020) Multiple mechanisms regulate H3 acetylation of enhancers in response to thyroid hormone. PLoS Genet 16(5):e1008770. https://doi.org/10.1371/journal.pgen.1008770
21. Smith CL, O'Malley BW (2004) Coregulator function: a key to understanding tissue specificity of selective receptor modulators. Endocr Rev 25(1):45–71. https://doi.org/10.1210/er.2003-0023
22. Lee DY, Teyssier C, Strahl BD, Stallcup MR (2005) Role of protein methylation in regulation of transcription. Endocr Rev 26(2):147–170. https://doi.org/10.1210/er.2004-0008
23. Lee YH, Stallcup MR (2009) Minireview: protein arginine methylation of nonhistone proteins in transcriptional regulation. Mol Endocrinol 23(4):425–433. https://doi.org/10.1210/me.2008-0380
24. Choi HK, Choi KC, Oh SY, Kang HB, Lee YH, Haam S et al (2007) The functional role of the CARM1-SNF5 complex and its associated HMT activity in transcriptional activation by thyroid hormone receptor. Exp Mol Med 39(4):544–555. https://doi.org/10.1038/emm.2007.60
25. Becker PB, Horz W (2002) ATP-dependent nucleosome remodeling. Annu Rev Biochem 71:247–273. https://doi.org/10.1146/annurev.biochem.71.110601.135400
26. Heimeier RA, Hsia VS, Shi YB (2008) Participation of Brahma-related gene 1 (BRG1)-associated factor 57 and BRG1-containing chromatin remodeling complexes in thyroid hormone-dependent gene activation during vertebrate development. Mol Endocrinol 22(5):1065–1077. https://doi.org/10.1210/me.2007-0492
27. Gillis NE, Boyd JR, Tomczak JA, Frietze S, Carr FE (2022) Thyroid hormone dependent transcriptional programming by TRbeta requires SWI/SNF chromatin remodelers. Nucleic Acids Res 50(3):1382–1395. https://doi.org/10.1093/nar/gkab1287
28. Fondell JD (2013) The mediator complex in thyroid hormone receptor action. Biochim Biophys Acta 1830(7):3867–3875. https://doi.org/10.1016/j.bbagen.2012.02.012
29. Femia MR, Evans RM, Zhang J, Sun X, Lebegue CJ, Roggero VR et al (2020) Mediator subunit MED1 modulates intranuclear dynamics of the thyroid hormone receptor. J Cell Biochem 121(4):2909–2926. https://doi.org/10.1002/jcb.29532
30. Sharma D, Fondell JD (2002) Ordered recruitment of histone acetyltransferases and the TRAP/mediator complex to thyroid hormone-responsive promoters in vivo. Proc

Natl Acad Sci USA 99(12):7934–7939. https://doi.org/10.1073/pnas.122004799

31. Metivier R, Reid G, Gannon F (2006) Transcription in four dimensions: nuclear receptor-directed initiation of gene expression. EMBO Rep 7(2):161–167. https://doi.org/10.1038/sj.embor.7400626
32. Perissi V, Aggarwal A, Glass CK, Rose DW, Rosenfeld MG (2004) A corepressor/coactivator exchange complex required for transcriptional activation by nuclear receptors and other regulated transcription factors. Cell 116(4):511–526. https://doi.org/10.1016/s0092-8674(04)00133-3
33. Zhang J, Hu X, Lazar MA (1999) A novel role for helix 12 of retinoid X receptor in regulating repression. Mol Cell Biol 19(9):6448–6457. https://doi.org/10.1128/MCB.19.9.6448
34. Heinzel T, Lavinsky RM, Mullen TM, Soderstrom M, Laherty CD, Torchia J et al (1997) A complex containing N-CoR, mSin3 and histone deacetylase mediates transcriptional repression. Nature 387(6628):43–48. https://doi.org/10.1038/387043a0
35. Huang EY, Zhang J, Miska EA, Guenther MG, Kouzarides T, Lazar MA (2000) Nuclear receptor corepressors partner with class II histone deacetylases in a Sin3-independent repression pathway. Genes Dev 14(1):45–54
36. Guenther MG, Barak O, Lazar MA (2001) The SMRT and N-CoR corepressors are activating cofactors for histone deacetylase 3. Mol Cell Biol 21(18):6091–6101. https://doi.org/10.1128/MCB.21.18.6091-6101.2001
37. Ishizuka T, Lazar MA (2005) The nuclear receptor corepressor deacetylase activating domain is essential for repression by thyroid hormone receptor. Mol Endocrinol 19(6):1443–1451. https://doi.org/10.1210/me.2005-0009
38. Astapova I, Hollenberg AN (2013) The in vivo role of nuclear receptor corepressors in thyroid hormone action. Biochim Biophys Acta 1830(7):3876–3881. https://doi.org/10.1016/j.bbagen.2012.07.001
39. Mendoza A, Astapova I, Shimizu H, Gallop MR, Al-Sowaimel L, MacGowan SMD et al (2017) NCoR1-independent mechanism plays a role in the action of the unliganded thyroid hormone receptor. Proc Natl Acad Sci USA 114(40):E8458–E8E67. https://doi.org/10.1073/pnas.1706917114
40. Grontved L, Waterfall JJ, Kim DW, Baek S, Sung MH, Zhao L et al (2015) Transcriptional activation by the thyroid hormone receptor through ligand-dependent receptor recruitment and chromatin remodelling. Nat Commun 6:7048. https://doi.org/10.1038/ncomms8048
41. Shabtai Y, Nagaraj NK, Batmanov K, Cho YW, Guan Y, Jiang C et al (2021) A coregulator shift, rather than the canonical switch, underlies thyroid hormone action in the liver. Genes Dev 35(5–6):367–378. https://doi.org/10.1101/gad.345686.120
42. Cho YW, Fu Y, Huang CJ, Wu X, Ng L, Kelley KA et al (2023) Thyroid hormone-regulated chromatin landscape and transcriptional sensitivity of the pituitary gland. Commun Biol 6(1):1253. https://doi.org/10.1038/s42003-023-05546-y
43. Shibusawa N, Hashimoto K, Nikrodhanond AA, Liberman MC, Applebury ML, Liao XH et al (2003) Thyroid hormone action in the absence of thyroid hormone receptor DNA-binding in vivo. J Clin Invest 112(4):588–597. https://doi.org/10.1172/JCI18377
44. Hones GS, Rakov H, Logan J, Liao XH, Werbenko E, Pollard AS et al (2017) Noncanonical thyroid hormone signaling mediates cardiometabolic effects in vivo. Proc Natl Acad Sci USA 114(52):E11323–E11E32. https://doi.org/10.1073/pnas.1706801115
45. Ramadoss P, Abraham BJ, Tsai L, Zhou Y, Costa-e-Sousa RH, Ye F et al (2014) Novel mechanism of positive versus negative regulation by thyroid hormone receptor beta1 (TRbeta1) identified by genome-wide profiling of binding sites in mouse liver. J Biol Chem 289(3):1313–1328. https://doi.org/10.1074/jbc.M113.521450
46. Machado DS, Sabet A, Santiago LA, Sidhaye AR, Chiamolera MI, Ortiga-Carvalho TM et al (2009) A thyroid hormone receptor mutation that dissociates thyroid hormone regulation of gene expression in vivo. Proc Natl Acad Sci USA 106(23):9441–9446. https://doi.org/10.1073/pnas.0903227106
47. You SH, Liao X, Weiss RE, Lazar MA (2010) The interaction between nuclear receptor corepressor and histone deacetylase 3 regulates both positive and negative thyroid hormone action in vivo. Mol Endocrinol 24(7):1359–1367. https://doi.org/10.1210/me.2009-0501
48. Weiss RE, Xu J, Ning G, Pohlenz J, O'Malley BW, Refetoff S (1999) Mice deficient in the steroid receptor co-activator 1 (SRC-1) are resistant to thyroid hormone. EMBO J 18(7):1900–1904. https://doi.org/10.1093/emboj/18.7.1900
49. Vella KR, Ramadoss P, Costa ESRH, Astapova I, Ye FD, Holtz KA et al (2014) Thyroid hormone signaling in vivo requires a

balance between coactivators and corepressors. Mol Cell Biol 34(9):1564–1575. https://doi.org/10.1128/MCB.00129-14

50. Perez P, Schonthal A, Aranda A (1993) Repression of c-fos gene expression by thyroid hormone and retinoic acid receptors. J Biol Chem 268(31):23538–23543
51. Lasa M, Gil-Araujo B, Palafox M, Aranda A (2010) Thyroid hormone antagonizes tumor necrosis factor-alpha signaling in pituitary cells through the induction of dual specificity phosphatase 1. Mol Endocrinol 24(2):412–422. https://doi.org/10.1210/me.2009-0298
52. Mendez-Pertuz M, Sanchez-Pacheco A, Aranda A (2003) The thyroid hormone receptor antagonizes CREB-mediated transcription. EMBO J 22(12):3102–3112. https://doi.org/10.1093/emboj/cdg295
53. Alonso-Merino E, Martin Orozco R, Ruiz-Llorente L, Martinez-Iglesias OA, Velasco-Martin JP, Montero-Pedrazuela A et al (2016) Thyroid hormones inhibit TGF-beta signaling and attenuate fibrotic responses. Proc Natl Acad Sci USA 113(24):E3451–E3460. https://doi.org/10.1073/pnas.1506113113
54. Contreras-Jurado C, Alonso-Merino E, Saiz-Ladera C, Valino AJ, Regadera J, Alemany S et al (2016) The thyroid hormone receptors inhibit hepatic Interleukin-6 signaling during Endotoxemia. Sci Rep 6:30990. https://doi.org/10.1038/srep30990
55. Sasaki S, Matsushita A, Kuroda G, Nakamura HM, Oki Y, Suda T (2018) The mechanism of negative transcriptional regulation by thyroid hormone: lessons from the thyrotropin beta subunit Gene. Vitam Horm 106:97–127. https://doi.org/10.1016/bs.vh.2017.06.006
56. Cao X, Kambe F, Moeller LC, Refetoff S, Seo H (2005) Thyroid hormone induces rapid activation of Akt/protein kinase B-mammalian target of rapamycin-p70S6K cascade through phosphatidylinositol 3-kinase in human fibroblasts. Mol Endocrinol 19(1):102–112. https://doi.org/10.1210/me.2004-0093
57. Furuya F, Hanover JA, Cheng SY (2006) Activation of phosphatidylinositol 3-kinase signaling by a mutant thyroid hormone beta receptor. Proc Natl Acad Sci USA 103(6):1780–1785. https://doi.org/10.1073/pnas.0510849103
58. Hiroi Y, Kim HH, Ying H, Furuya F, Huang Z, Simoncini T et al (2006) Rapid nongenomic actions of thyroid hormone. Proc Natl Acad Sci USA 103(38):14104–14109. https://doi.org/10.1073/pnas.0601600103
59. Storey NM, Gentile S, Ullah H, Russo A, Muessel M, Erxleben C et al (2006) Rapid signaling at the plasma membrane by a nuclear receptor for thyroid hormone. Proc Natl Acad Sci USA 103(13):5197–5201. https://doi.org/10.1073/pnas.0600089103
60. Moeller LC, Dumitrescu AM, Refetoff S (2005) Cytosolic action of thyroid hormone leads to induction of hypoxia-inducible factor-1alpha and glycolytic genes. Mol Endocrinol 19(12):2955–2963. https://doi.org/10.1210/me.2004-0542
61. Martin NP, Fernandez M, de Velasco E, Mizuno F, Scappini EL, Gloss B, Erxleben C et al (2014) A rapid cytoplasmic mechanism for PI3 kinase regulation by the nuclear thyroid hormone receptor, TRbeta, and genetic evidence for its role in the maturation of mouse hippocampal synapses in vivo. Endocrinology 155(9):3713–3724. https://doi.org/10.1210/en.2013-2058
62. Hones GS, Kerp H, Hoppe C, Kowalczyk M, Zwanziger D, Baba HA et al (2022) Canonical thyroid hormone receptor beta action stimulates hepatocyte proliferation in male mice. Endocrinology 163(3). https://doi.org/10.1210/endocr/bqac003
63. Carazo A, Levin J, Casas F, Seyer P, Grandemange S, Busson M et al (2012) Protein sequences involved in the mitochondrial import of the 3,5,3′-L-triiodothyronine receptor p 43. J Cell Physiol 227(12):3768–3777. https://doi.org/10.1002/jcp.24085
64. Wrutniak-Cabello C, Casas F, Cabello G (2017) Mitochondrial T3 receptor and targets. Mol Cell Endocrinol 458:112–120. https://doi.org/10.1016/j.mce.2017.01.054
65. Wrutniak C, Cassar-Malek I, Marchal S, Rascle A, Heusser S, Keller JM et al (1995) A 43-kDa protein related to c-Erb a alpha 1 is located in the mitochondrial matrix of rat liver. J Biol Chem 270(27):16347–16354. https://doi.org/10.1074/jbc.270.27.16347
66. Casas F, Daury L, Grandemange S, Busson M, Seyer P, Hatier R et al (2003) Endocrine regulation of mitochondrial activity: involvement of truncated RXRalpha and c-Erb Aalpha1 proteins. FASEB J 17(3):426–436. https://doi.org/10.1096/fj.02-0732com
67. Morrish F, Buroker NE, Ge M, Ning XH, Lopez-Guisa J, Hockenbery D et al (2006) Thyroid hormone receptor isoforms localize to cardiac mitochondrial matrix with potential for binding to receptor elements on mtDNA. Mitochondrion 6(3):143–148. https://doi.org/10.1016/j.mito.2006.04.002
68. Bergh JJ, Lin HY, Lansing L, Mohamed SN, Davis FB, Mousa S et al (2005) Integrin alphaVbeta3 contains a cell surface receptor site for thyroid hormone that is linked to

activation of mitogen-activated protein kinase and induction of angiogenesis. Endocrinology 146(7):2864–2871. https://doi.org/10.1210/en.2005-0102

69. Davis PJ, Goglia F, Leonard JL (2016) Nongenomic actions of thyroid hormone. Nat Rev Endocrinol 12(2):111–121. https://doi.org/10.1038/nrendo.2015.205
70. Davis PJ, Leonard JL, Lin HY, Leinung M, Mousa SA (2018) Molecular basis of nongenomic actions of thyroid hormone. Vitam Horm 106:67–96. https://doi.org/10.1016/bs.vh.2017.06.001
71. Freindorf M, Furlani TR, Kong J, Cody V, Davis FB, Davis PJ (2012) Combined QM/MM study of thyroid and steroid hormone analogue interactions with alphavbeta3 integrin. J Biomed Biotechnol 2012:959057. https://doi.org/10.1155/2012/959057
72. Lin HY, Sun M, Tang HY, Lin C, Luidens MK, Mousa SA et al (2009) L-thyroxine vs. 3,5,3′-triiodo-L-thyronine and cell proliferation: activation of mitogen-activated protein kinase and phosphatidylinositol 3-kinase. Am J Physiol Cell Physiol 296(5): C980–C991. https://doi.org/10.1152/ajpcell.00305.2008
73. Davis PJ, Mousa SA, Lin HY (2021) Nongenomic actions of thyroid hormone: the integrin component. Physiol Rev 101(1):319–352. https://doi.org/10.1152/physrev.00038.2019
74. Rangaswami H, Schwappacher R, Marathe N, Zhuang S, Casteel DE, Haas B et al (2010) Cyclic GMP and protein kinase G control a Src-containing mechanosome in osteoblasts. Sci Signal 3(153):ra91. https://doi.org/10.1126/scisignal.2001423
75. Farwell AP, Dubord-Tomasetti SA, Pietrzykowski AZ, Leonard JL (2006) Dynamic nongenomic actions of thyroid hormone in the developing rat brain. Endocrinology 147(5): 2567–2574. https://doi.org/10.1210/en.2005-1272
76. Farwell AP, Dubord-Tomasetti SA, Pietrzykowski AZ, Stachelek SJ, Leonard JL (2005) Regulation of cerebellar neuronal migration and neurite outgrowth by thyroxine and 3,3′,5′-triiodothyronine. Brain Res Dev Brain Res 154(1):121–135. https://doi.org/10.1016/j.devbrainres.2004.07.016
77. Ariyani W, Miyazaki W, Amano I, Koibuchi N (2022) Involvement of integrin alphavbeta3 in thyroid hormone-induced dendritogenesis. Front Endocrinol (Lausanne) 13:938596. https://doi.org/10.3389/fendo.2022.938596

Part II

Thiroid Hormones

Chapter 3

A Sensitive Radioimmunoassay for T3 and T4 Determinations in Plasma and Tissues

Ana Montero-Pedrazuela, Carmen Grijota-Martínez, Ana Guadaño-Ferraz, and Soledad Bárez-López

Abstract

This chapter details protocols for determining plasma thyroid hormone (TH) levels and tissue TH content by competitive radioimmunoassays (RIAs). These protocols include: an initial test of the chromatographic performance, isotopic labeling to produce high activity ^{125}I-T3 and ^{125}I-T4, free iodide estimation of the labeled products, purification of tracers from iodide by paper electrophoresis, extraction of THs from plasma and tissue samples, and the RIA procedures. The RIA involves the competition between radioactive labeled and unlabeled hormones for specific antibody binding, and due to its high sensitivity is capable of detecting a minimum of 2.5 pg of T4 and 0.4 pg of T3.

Drs. María Jesús Obregon and Gabriella Morreale de Escobar improved these protocols in the 1970s, enhancing sensitivity and accuracy. Their improvements enabled the detection of TH content in tissues, providing crucial insights into maternal THs role in fetal development and the importance of iodine intake during gestation. Their research also facilitated the early detection of congenital hypothyroidism, preventing neurological impairments in newborns. Internationally and in Spain, their contributions are widely acknowledged, leading to substantial public health impacts, including the implementation of nationwide neonatal screening programs. Despite progress, thyroid diseases remain prevalent, underscoring the need for continued research into thyroid physiology and associated disorders, employing highly sensitive techniques like the RIA outlined herein that we continue using in our research.

Key words Radioimmunoassay, Thyroid hormones, Triiodothyronine, T3, Thyroxine, T4, Radioactivity, Hypothyroidism, Iodide, Chromatography

1 Introduction

Competitive radioimmunoassays (RIAs) involve the competition between non-labeled and isotopically labeled hormones (or other substances) for binding to specific antibodies (Fig. 1). To perform thyroid hormones (THs) determinations, a known amount of radiolabeled TH is added to the sample, along with the specific antibody that recognizes both the radiolabeled and the endogenous TH. As the radiolabeled TH competes with the endogenous for

Constanza Contreras-Jurado (ed.), *Thyroid Hormones: Methods and Protocols*, Methods in Molecular Biology, vol. 2876, https://doi.org/10.1007/978-1-0716-4252-8_3,

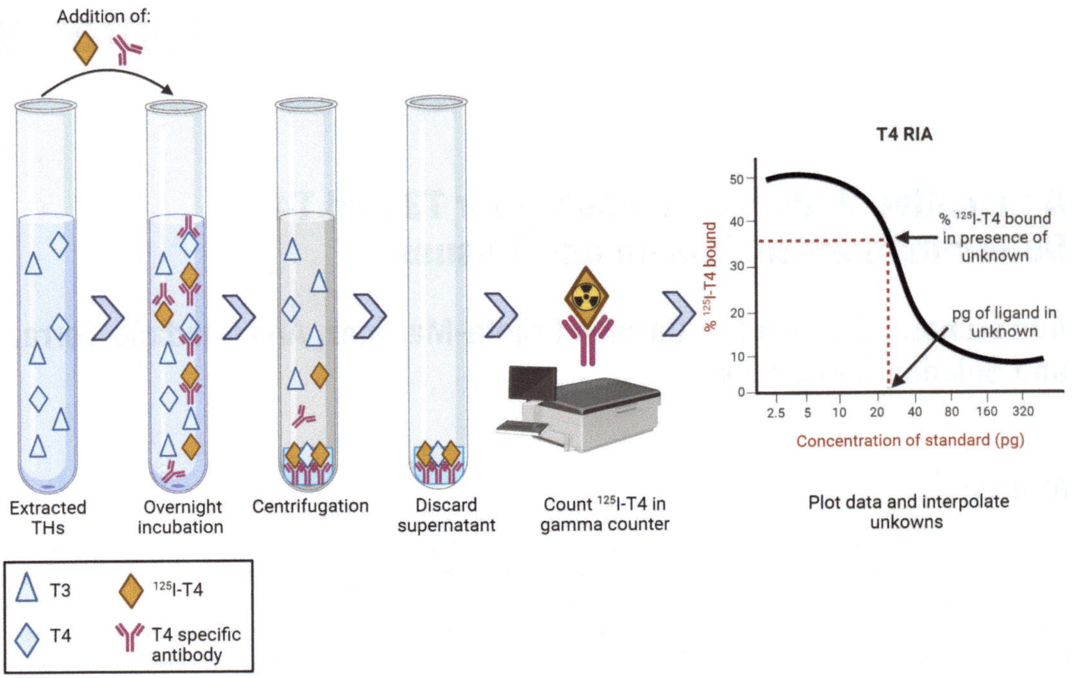

Fig. 1 Radioimmunoassay principles. Schematic representation of the main steps of the RIA procedure using RIA for T4 as an example. After purifying and extracting THs from the tissue sample, an aliquot of the extract is combined with the specific antibody and the isotopically labeled hormone (^{125}I-T4 for this example). During overnight incubation, there is a competition between non-labeled and isotopically labeled hormones for binding to specific antibodies and the complex antibody-hormone (either labeled or non-labeled) is formed. The bound hormone is then precipitated by centrifugation, and, after discarding the supernatant, the radioactivity present in the pellet is measured by means of a gamma counter to obtain the T4 content of the samples. A standard curve is then generated using known amounts of T4 (X axis) and the corresponding percentages (% bound, Y axis) of associated ^{125}I-T4. By interpolating the % of ^{125}I-T4 found in unknown samples, the concentration of T4 present in those samples can be estimated. Figure generated with BioRender.com

binding sites on the antibody, the amount of radiolabeled hormone bound to the antibody inversely correlates with the concentration of TH present in the sample. In addition, a standard curve is generated using increasing known amounts of non-labeled hormone, along with a fixed amount of labeled hormone and a fixed amount of specific antiserum. The samples containing the unknown endogenous amounts of TH are combined with the fixed amounts of radiolabeled hormone and antiserum, as indicated for the standard curve (Fig. 1). The antisera-bound hormone is then precipitated and the supernatant containing the unbound hormone is subsequently discarded. The bound radioactivity is then measured. The standard curve is used to establish the equivalence between picograms (pg) of non-labeled hormone and counts per minute (cpm) which corresponds to the percentage of radioactive TH bound to antibody. By plotting this curve, the radioactive measurements of the samples can be interpolated to determine the unknown amount (pg) of TH present in each sample (Fig. 1).

In this chapter, we will detail the protocols established by Drs. María Jesús Obregon and Gabriella Morreale de Escobar in the 1970s for the determination of the THs: T3 and T4 (3′,3,5-triiodothyronine and 3′,5′,3,5-tetraiodothyronine or thyroxine) in plasma and tissues using competitive RIAs. Despite the years, we continue to routinely use these protocols today at our laboratory for exploring many physiological and pathophysiological questions. These protocols for TH determinations encompass several crucial steps: firstly, conducting an initial chromatographic test to assess the optimal separation of T4, T3 and iodide; next, performing isotopic labeling to generate high activity ^{125}I-T3 and ^{125}I-T4; then, free iodide estimation of the labeled products; subsequently, purification of tracers from iodide by paper electrophoresis to estimate the recovery of endogenous THs after the extraction procedure; then, extraction of THs from plasma and tissue samples; and finally conducting the RIA procedure.

The protocols established by Drs. Morreale de Escobar and Obregon were based on a highly sensitive RIA to measure free T3 in human serum originally developed by Weeke and Orskov in 1975 that was capable of detecting as low as 1 pg of T3 in serum samples [1]. Drs. Obregon and Morreale de Escobar introduced important modifications resulting in improvements in the synthesis of high-activity radiolabeled hormones (that enhances the accuracy and sensitivity), in the method for extracting THs from tissue samples (which is particularly challenging in high lipid content tissues such as the brain due to its impact on the RIA reaction), and in approaches that avoid cross-reactivity of THs with other TH derivatives [2, 3].

Drs. Obregon and Morreale de Escobar introduced modifications in the method of products separation following radioactive labeling (^{125}I-Iodide, ^{125}I-T4, and ^{25}I-T3) [4] by using a paper chromatography system that proved to be cheaper and more user-friendly. In addition, while the initial RIA methods described by Weeke and Orskov involved the use of wick chromatography to separate bound (antigen-antibody complex) from free hormones, a very sensitive but expensive and time-consuming method, Drs. Morreale de Escobar and Obregon optimized the use of polyethylene glycol 6000 (PEG) as a precipitating agent. Alongside serum acting as a carrier of gamma-globulins to facilitate the process, this approach resulted to be cheaper and quicker while preserving the required sensitivity.

The development of this technique by these two exceptional scientists allowed the formulation of some of the fundamental principles of modern endocrinology. For instance, the possibility of accurately detecting very small amounts of THs in tissues, which had not been possible prior to the refinement of this protocol, facilitated the detection of THs content in fetal tissues before the onset of the fetal thyroid gland's functionality. This seminal finding supported the fact that there is a transfer of THs from the mother to the fetus during gestation influencing fetal brain development [5]. In follow-up studies using this RIA protocol, Drs. Obregon and Morreale de Escobar proved the importance of maintaining an adequate dietary iodine intake during gestation to ensure the proper synthesis of THs and their transfer to the fetus [6]. Consequently, the widespread adoption of administering iodine supplements to pregnant women, to avoid lower levels of THs in plasma, became a common practice to mitigate fetal developmental alterations. Moreover, these findings, alongside the identification of brain defects in TH-dependent functions observed in premature infants at 6–7 months of gestation [7], support the importance of maternal THs transfer to the fetus during the later stages of pregnancy and therefore the supplementation of premature infants with THs.

In a series of preclinical studies using the RIA as a pivotal experimental approach, Gabriella Morreale de Escobar and colleagues proved the importance of early detection of congenital hypothyroidism to prevent the associated neurological impairments. This motivated Dr. Morreale de Escobar and coworkers to launch a clinical pilot study in 1976 to screen newborns for congenital hypothyroidism, using T4 and TSH [8], representing another significant contribution facilitated by highly sensitive RIAs. This program gradually expanded worldwide, enabling early diagnosis and treatment of neonates with TH, and preventing mental disability in around 150 children annually in Spain (frequency 1 per 2000–4000 births).

In addition, Drs. Gabriella Morreale de Escobar and Francisco Escobar del Rey, her husband, contributed to implementing a National health regulation on the necessity of iodizing common salt to ensure sufficient iodine intake among the population, thus preventing the potential problems caused by iodine deficiency.

The legacy of Dr. Gabriella Morreale de Escobar and her colleagues, such as Dr. María Jesús Obregon, has been widely recognized within the Spanish and international scientific community, receiving the First Reina Sofia Prize for the Prevention of Mental Disabilities in 1983. Notably, one of the most prestigious national research awards in Spain, the National Prize for Young Researchers in Medicine and Health Sciences, is named after Gabriella Morreale.

Despite the advances in this field, thyroid diseases are still the most common endocrine disorder worldwide, and the incidence of thyroid conditions has increased over the past few decades. In addition to genetic diseases, later evidence points to endocrine disruptors as potential thyroid disruptors, impacting the synthesis and action of T3 and T4 [9]. To comprehensively address these and other issues, it is crucial to determine TH levels and content in both plasma and tissues, and this can be achieved using specific RIAs.

Here we recognize the invaluable contribution that this analytical method has made and will continue to make in enhancing our understanding of thyroid physiology and the mechanisms mediating THs availability, metabolism, and action in different tissues.

2 Materials

2.1 Testing Chromatographic Separation

1. Chromatographic tank specific for testing (15 cm width × 34 cm length × 8.5 cm depth, with a glass rod horizontally attached at 5 cm high and 4 cm from one end to support the paper strips, a loose rod to keep the paper strips in place at the bottom of the tank and a tight lid; Fig. 2).
2. Cellulose chromatography paper (Whatman 3MM), 2.5 cm width, 50 cm long strips.
3. Carrier: 8 mg thyroxine, 8 mg triiodothyronine, 2 mg potassium iodide in 2 mL of methanol:2 N ammonia (3:1).
4. Chromatographic solvent (BEA): Butanol: Ethanol: 0.5 N Ammonia (5:1:2). Prepare a minimum of 500 mL.
5. Developer solution 1: 0.5% Cl_2Pd in 0.1 N HCl (dilute 1:10 with water to obtain the working solution).
6. Developer solution 2: 1% aminoantipyrine in 2% Na_2CO_3.
7. Developer solution 3: 2% $K_3Fe(CN)_6$ in dH_2O.

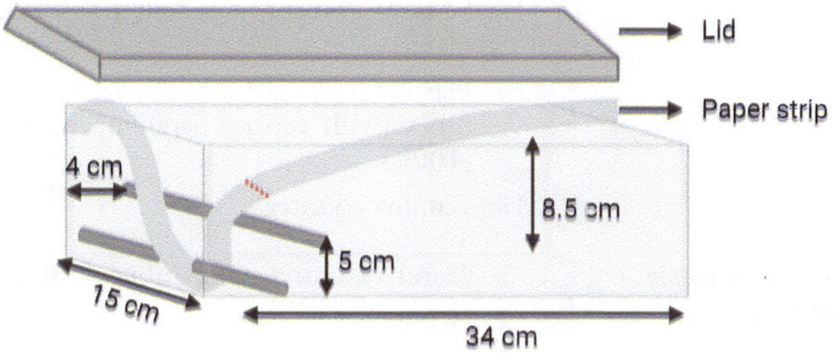

Fig. 2 Chromatographic tank. Schematic representation of the chromatographic tank with its inner measures and the position of the rods (dark gray cylinders). The positioning of the chromatographic paper strip (medium gray) is also visualized. Note the red dotted line, where the labeling reaction is to be placed. Figure generated with BioRender.com

2.2 Synthesis of High Specific Activity ^{125}I-T3 and ^{125}I-T4

1. Equipped radiation room (Geiger counter, lead-covered refrigerator, and fume hood with specific filters for radioactive nascent $^{125}I\text{-}I_2$ and disposal containers).
2. Personal protection equipment (personal dosimetry device, disposable lab coat, over-sleeves, glovers, shoe-covers, and appropriate mask).
3. ^{125}I-Iodide-Radionuclide, high specific activity (~17 Ci/mg) in 10^{-5} M NaOH (pH 8–11), carrier-free (such as Revvity, NEZ033A).
4. Phosphate buffer (PB) 0.5 M and 0.05 M.
5. Substrate for ^{125}I-T4 radioactive labeling: 100–200 μg/mL T3 in 0.04 N NaOH.
6. Substrate for ^{125}I-T3 radioactive labeling: 100–200 μg/mL T2 (3, 5- di-iodo-thyronine) in 0.04 N NaOH.
7. One 1.5 mL tube (with screw cap) and lead container per reaction.
8. Cellulose chromatography paper (Whatman 3MM), 2.5 cm width, 30 cm long strips. One strip per reaction.
9. 2.5 mg/mL Chloramine T in 0.05 M PB. Prepare the solution immediately before use (*see* **Note 1**).
10. 20.8 mg/mL Metabisulfite in 0.05 M PB. Prepare the solution immediately before use (*see* **Note 1**).
11. Chronometer, vortex, pipettes, and filter pipette tips.
12. Chromatography tank specific for radioactive material (15 cm width × 34 cm length × 8.5 cm depth, with a glass rod horizontally attached at 5 cm high and 4 cm from one end to support the paper strips, a loose rod to keep the paper strips in place at the bottom of the tank and a tight lid; Fig. 2) and BEA solvent.
13. Scissors, forceps, a pencil, and a rule (specific for radioactive material).
14. Plastic tubes (~12 mm diameter, 24 tubes per substrate), glass tubes (with cap or parafilm; 1 tube per substrate), ethanol (100%).
15. Gamma counter.

2.3 Free Iodide Estimation

1. Gamma counter-compatible plastic tubes, ethanol (100%).
2. Horse or calf serum.
3. Gamma counter.
4. 10% Trichloroacetic acid (TCA).
5. Vortex, refrigerated centrifuge, and vacuum pump with recipient to collect radioactive material.

2.4 Purification of Tracers from Iodide by Paper Electrophoresis

1. Cellulose chromatography paper (Whatman 3MM), 2.5 × 23 cm strips.
2. Chronometer, pipettes and filter pipette tips.
3. Paper electrophoretic chamber (containing electrodes, buffer reservoirs, a platform for supporting paper, and a transparent insulating cover for support) and a power pack to provide a current of 400 V and 10–50 mA.
4. 100 mM Ammonium acetate buffer pH 6.8.
5. Scissors and forceps.
6. Glass tubes for elution, ethanol (100%).
7. Screw cap 1.5 mL tube.

2.5 T3 and T4 Extraction from Plasma Samples

1. Pipettes and pipette tips, 1.5 mL tubes.
2. Vortex, table-top centrifuge, fume hood.
3. Methanol.
4. Gamma counter-compatible tubes.
5. Gamma counter.
6. Speed-vac.
7. Stock RIA buffer (SR buffer): 0.04 M PB pH 8, 0.6 mM Merthiolate.
8. RIA buffer A: SR buffer with 0.2% bovine serum albumin (BSA; *see* **Note 2**).

2.6 Preparation of T3 and T4 Extraction Columns for Tissue Samples

1. AG 1-X2 Anion Exchange Resin, analytical grade, 200–400 mesh, chloride form (BioRad #1401251).
2. Acetic acid.
3. Syringes (2 mL) and compatible tall rack.
4. Cellulose chromatography paper (Whatman 3MM) and paper drill.

2.7 T3 and T4 Extraction from Tissue Samples

1. Tissue homogenizer.
2. Methanol.
3. Pipettes and pipette tips.
4. Vortex, table-top centrifuge, fume hood.
5. Plastic tubes (10 mL, chloroform-resistant).
6. Chloroform.
7. 0.05% Cl_2Ca.
8. Extraction columns prepared in the previous step.
9. Solution A: 11.5 mL acetic acid in 1 L dH_2O.
10. Solution B: 16.4 g anhydrous sodium acetate in 1 L dH_2O.

11. Acetate buffer pH 7: 1.25 mL of solution A in 1 L of solution B.
12. Acetate buffer pH 4: 180 mL of solution B in 1 L of solution A.
13. Ethanol (100%).
14. Acetic acid (1%, 35%, 70%).
15. Gamma counter-compatible tubes.
16. Gamma counter, speed-vac.
17. SR buffer: 0.04 M PB pH 8, 0.6 mM Merthiolate.
18. RIA buffer A: SR buffer with 0.2% BSA (*see* **Note 2**).

2.8 Radioimmunoassays for T3 and T4 Determinations

1. T3 standard curve stock solution: 4 μg T3 per mL in NaOH 0.05 N.
2. T3 standard curve working solution: 5 ng T3 per mL in NaOH 0.05 N.
3. T4 standard curve stock solution: 10 μg T4 per mL in NaOH 0.05 N.
4. T4 standard curve working solution: 64 ng T4 per mL in NaOH 0.05 N.
5. RIA tubes (specifications: flat bottom glass tubes, 80 mm length, 8 mm inner diameter) and grids.
6. Pipettes and pipette tips.
7. RIA buffer A: SR buffer with 0.2% BSA for T3 RIA (*see* **Note 2**).
8. RIA buffer B: SR buffer with 0.05% BSA for T4 RIA (*see* **Note 2**).
9. Anti-T3 and anti-T4 antisera (*see* **Note 3**).
10. 10 mg/mL Cysteine.
11. Polyethylene glycol 6000 (PEG).
12. Calf serum (*see* **Note 4**).
13. SR buffer.
14. Liquid dispenser.
15. Vortex, refrigerated high-capacity centrifuge.
16. Vacuum pump equipped with a container for radioactive liquid waste.
17. Gamma counter.

3 Methods

Preventing Cross-Contamination in RIA Procedures

For all the protocol is of utmost importance to avoid cross-contamination of the material and buffers with iodide and/or THs. It is essential to use specific glass and plastic materials only for RIA procedures and to manipulate carefully all buffers and solutions. Be particularly vigilant when using the carrier for chromatography testing (Subheading 3.1) and when preparing the standard curves for T3 and T4 RIAs (Subheadings 3.8 and 3.9). It is advisable to prepare these experiments in a separate area.

3.1 Testing Chromatographic Separation

The aim of this test is to verify the correct separation of iodothyronines for the recovery of the labeled THs after the isotopic labeling (next step). The separation depends on the maturity of the BEA buffer (meaning its saturation with ammonia), on environmental conditions, and on the running time.

1. Prepare a strip of chromatography paper: make the folds to fit the chromatographic tank, leaving the strip well taut and create a reference point by making a mark 4 cm above the fixed rod (Fig. 2).
2. Inside a switched-on hood, apply 25 μL of the carrier throughout that line (keep as thin as possible). Allow to dry slightly, place it in the chromatographic tank and add the BEA solvent to the bottom of the container carefully (200 mL approx.).
3. Close the tank and tilt about 45°. Close the hood and let the chromatography run for about 16–17 h. Register the running time.
4. The next day, take the strips out and allow them to dry by hanging them from the switched-on hood.
5. Once dry, spray Developer solution 1 to detect iodide on the middle part of the strip (where the iodide is located) until it develops a brownish-gray color. Mark the developed area with a pencil (the color will disappear after applying the Developer basic solution 2).
6. Then, spray Developer solution 2 over the entire strip, to achieve a basic pH.
7. Finally, spray Developer solution 3 onto the strip, revealing the iodothyronines with a pink color, above the iodide region. Notice that T4 appears with a deeper pink color and closer to iodide, while T3 manifests a subtle orange tone, positioned towards the front of the strip.
8. If the separation is not optimal, adjust the running time and/or prepare a new BEA solvent.

3.2 Synthesis of High Specific Activity ^{125}I-T3 and ^{125}I-T4

Follow the local radiation protection unit instructions for safety including radiation protection courses and training to handle radioactive materials. Always use suitable personal protective equipment (PPE), including gloves (two pairs), safety glasses, disposable lab coat, over-sleeve, shoe-covers and appropriate face mask, and personal dosimetry device. Store and manipulate ^{125}I source vials and solutions in lead containers at all times. Perform the reaction inside a proper fume hood behind lead glass shields. Use radiation-detection equipment (e.g., Geiger counter) to detect and prevent the spread of contamination. Check gloves, forearms, and other PPE, as well as working areas, for contamination frequently. Place radioactive waste materials in appropriate containers. Clearly label work areas, materials, and containers.

1. Prepare the substrates: place 50 μL of PB 0.5 M and 1 μg of the substrate per mCi to be used in the reaction (T2 (3,5-diiodothyronine) for T3 labeling and T3 for T4 labeling) in a screw cap 1.5 mL tube. Scale substrate and radionuclide when necessary, but always use 50 μL of PB 0.5 M to tampon the solution (*see* **Note 5**).
2. Prepare the chromatographic paper strips (one per reaction): fold them to fit the chromatographic tank, create a reference point by making a mark 4 cm above the fixed rod (Fig. 2), and label the strips at both ends using a pencil to provide a clear identification and orientation throughout the chromatography process.
3. Cover all work area inside the hood with protective and absorbent paper to protect against spills. Locate in the hood the required materials and prepare the chronometer.
4. Dissolve Chloramine T and Metabisulfite in PB 0.05 M using vortex (*see* **Note 1**).
5. Place the ^{125}I radioactivity container in the hood and make sure that the liquid is at the bottom of the vial. Use the thread of the leaded container to open the lid (DO NOT touch the vial directly).
6. Add the required amount of ^{125}I (1 mCi per μg of the substrate, usually 0.5 to 2 mCi) to the tube containing the substrate and PB 0.5 M.
7. Initiate the reaction by adding 10 μL of Chloramine T (25 μg). Start the timer and vortex the mixture carefully. The reaction time is critical (*see* **Note 6**).
8. Stop the reaction at 30 s, adding 25 μL of Metabisulfite (520 μg). Mix by vortexing.
9. Repeat **steps 6** to **8** for each substrate.

10. Immediately after the reactions, pipet the entire labeling reaction at the marked line of the chromatography strip forming a narrow line (the strip should be placed outside the tank to avoid its contamination). If needed, apply half of the volume and let several minutes to dry before applying the rest (if the paper soaks excessively, it could tear when placing the strip onsite).
11. After the strip is dry (less than 10 min), place each strip inside the tank (not touching each other), add the BEA solvent (200 mL approx.), and close the lid tightly. Make sure each strip is well taut, so it will not touch the liquid during the running.
12. Tilt the tank about 45°. Close the hood and let the chromatography run for the estimated time after chromatography testing (*see* Subheading 3.1).
13. The next day, take out the strips and let them dry on the hood for about 30 min.
14. Mark each strip every 0.8 cm from the front of the strip (using an ad hoc ruler covered with film and pencil) up to 24 marks.
15. Cut the strip into 0.8 cm wide strips using ad hoc scissors and forceps, and put each one into neat tubes labeled with numbers 1 to 24.
16. Measure the radioactivity for each tube using the Geiger counter at a constant distance (e.g., 30 cm) and write down the counts per second (cps) for each strip.
17. Identify the strips forming the peak containing the labeled hormone (from the front: the first peak is T3, then T4, and lastly iodide; *see* **Note** 7).
18. Select the strips containing the labeled hormone, cut into smaller pieces, place them in a glass tube with 2–3 mL of ethanol, and close the tube to prevent evaporation. Place the tube(s) into a leaded container in the leaded fridge (4 °C) overnight to elute.
19. Tidy up the working area: check everything with the Geiger counter and collect all the radioactive or non-radioactive waste into the appropriate containers. Decontaminate any instrument or surface that has become contaminated. Clearly label contaminated glassware or equipment until it has been decontaminated.
20. The next day count 1 μL of each labeled hormone in the gamma counter. Calculate the incorporated activity taking into account that 2.2×10^6 dpm (disintegrations per minute) is equal to 1 μCi, and the efficiency of the gamma counter. The present protocol is optimized to achieve high specific activity

labeled products, which will yield the best results in the RIA by increasing sensitivity and accuracy (*see* **Note 8**).

21. The labeled hormones will be used to generate the standard curve for the radioimmunoassay (RIA) and also as internal recovery tracers during the extraction of the THs from plasma and tissues. Before using the labeled hormones as tracers, it is advisable to estimate the free iodide (*see* next section). This free iodide is derived mainly from spontaneous deiodination and increases with time after the labeling, depending on storage conditions, and stability of the labeled products.

3.3 Free Iodide Estimation

1. Prepare 3 gamma counter-compatible tubes per labeled hormone: add 100 μL of the horse or calf serum and ≈ 8000 cpm of the labeled hormone per tube (dilute the hormone previously using ethanol if necessary, limiting the volume of ethanol to ≤2 μL per tube).
2. Measure the initial cpm and let balance for 15 min.
3. Add 400 μL of 10% TCA, vortex, and centrifuge 15 min 4 °C 2500 rpm.
4. Absorb the liquid (to radioactive waste) and count the tubes again in a gamma counter.
5. Estimate the free iodide, which should be less than 10% when using the ^{125}I -labeled T3 or T4 as tracers (*see* Subheading 3.4).

3.4 Purification of Tracers from Iodide by Paper Electrophoresis

To be used as tracers, the free iodide must be eliminated from the labeled hormones. Even if the free iodide estimation is low, this process should be performed. It is advisable to purify the tracers close to the moment of the extraction process.

1. Prepare the strips of 3MM Whatman paper (one per hormone): draw a line about 6–8 cm from one end using a pencil and make two folds in the middle part to achieve curvature. Label the strips to provide a clear identification.
2. Calculate the amount of each labeled hormone needed as a tracer for the corresponding samples to add 1000–1500 total cpm per tube (*see* **Note 9**).
3. Apply the labeled hormone on the line of the appropriate strip (preferably outside the electrophoretic chamber to avoid contamination). If the volume exceeds 20 μL, apply it in two times, letting it evaporate between applications.
4. Pour the buffer (ammonium acetate 100 mM pH 6.8) into the chamber, and place the strips of paper, holding them so that they are stable, with the origin at the negative pole. Carefully, soak the strip with buffer, avoiding direct application of the buffer onto the hormone-containing area. Instead, apply the

buffer to the rest of the strip, allowing for the hormone area to be saturated through diffusion.

5. Cover the electrophoretic chamber and apply 10 mA for each strip. Let the electrophoresis run for 5 min.
6. Stop the electrophoresis and cut the area surrounding the pencil line (the iodide should have migrated, and the hormone will be purified).
7. Put the piece of paper into a glass tube with 500 μL of ethanol (or the volume needed depending on the amount of labeled hormone applied; *see* **Note 10**). Let it elute at least overnight at 4 °C.
8. Count the amount of radioactivity per 10 μL. The final concentration should be calculated to add 10–20 μL to each extraction tube (with 1000–1500 total cpm). Transfer the liquid to a fresh tube and store at 4 °C.

3.5 T3 and T4 Extraction from Plasma Samples

Blood is collected in heparinized tubes and processed into plasma (*see* **Note 11**). Store all samples at −80 °C until extraction. For this protocol we will measure T3 and T4 concentrations in all samples. Modify accordingly if only one of the hormones is to be quantified (*see* **Note 9**). It is also possible to measure T3 and T4 in cerebrospinal fluid samples using a modified version of this protocol.

1. Thaw plasma samples in ice. Use a starting plasma volume of at least 80 μL in Eppendorf tubes to ensure the extraction of adequate levels of THs facilitating accurate T3 and T4 determinations. Pooling samples from different animals is an option if the required starting material is insufficient, although individual samples are recommended.
2. Add 1000–1500 cpm of the T3 tracer (in a volume of 10–20 μL) in half of the samples. Mix by vortex and leave 5 min to allow for the ^{125}I-T3 to equilibrate with the endogenous T3. Add the same volume and cpms of the T3 tracer in 3 gamma counter-compatible tubes (T3 reference tubes) and keep them apart.
3. Add 1000–1500 cpm of the T4 tracer in a volume of 10–20 μL in the other half of the samples. Mix by vortex and leave 5 min to allow for the ^{125}I-T4 to equilibrate with the endogenous T4. Add the same volume and cpms of the T4 tracer in 3 gamma counter-compatible tubes (T4 reference tubes) and keep them apart.
4. Add 500 μL methanol to each sample. Mix by vortexing and centrifuge at 5000 rpm for 10 min at room temperature.
5. Transfer the supernatant to a gamma counter-compatible tube.

6. Add 500 μL methanol to the pellet obtained in **step 4**. Resuspend by vigorously vortexing and centrifuge at 5000 rpm for 10 min at room temperature.
7. Transfer the supernatant to the same gamma counter-compatible tube used in **step 5**.
8. Totally evaporate the methanol in the samples using a SpeedVac vacuum concentrator and resuspend in 310 μL of RIA buffer A by vortexing.
9. Count the T3 and T4 tracer cpms in the sample after extraction using a gamma counter and calculate the percentage of T3 and T4 recovery by comparing each sample with the T3 and T4 reference tubes prepared in **steps 2** and **3**. For samples containing the T3 tracer, estimate the T4 recovery by determining the mean value from all samples containing the T4 tracer, and vice versa.
10. Keep the extracted plasma samples at −20 °C until further use.

3.6 Preparation of T3 and T4 Extraction Columns for Tissue Samples

1. To prepare 75 extraction columns, transfer 40 g of AG 1-X2 Anion Exchange Resin to a jar and cover completely with 70% acetic acid in a fume hood. Cover with lid, stir by vigorously shaking, and allow to soak overnight.
2. Once the AG 1-X2 Anion Exchange Resin has settled, pour away the 70% acetic acid.
3. Add distilled water (dH_2O), and stir by vigorously shaking. Once the resin has settled, pour away the dH_2O and repeat this step four or five times until the acetic acid has washed away. Maintain the resin in dH_2O for no longer than 1 month to prevent contamination.
4. In the meantime, remove the plunger from 75 units of 2-mL syringes and place the syringes without the plunger in a tall rack.
5. Using a paper drill, cut 75 pieces of Whatman 3MM paper circles with approximately a diameter of 6 mm.
6. Place one Whatman 3MM paper circle at the bottom of each syringe covering the end of the barrel. Add a few μL of dH_2O to ensure that the Whatman 3MM paper circle stays precisely at the needle hub.
7. Add AG 1-X2 Anion Exchange Resin in dH_2O to the syringe until it reaches the 1.2 mL grading. Ensure no resin spills or leaks from the syringe, as any escape will interfere with the RIA. Keep the columns at 4 °C for not longer than 7 days.

3.7 T3 and T4 Extraction from Tissue Samples

The tissues of interest are harvested, weighted, and frozen in dry ice (*see* **Note 12**). Store all samples at −80 °C until extraction.

1. Homogenize frozen tissue samples using methanol and a tissue homogenizer, such as a Polytron homogenizer (*see* **Note 13**). For tissue samples up to 200 mg, use 500 μL methanol (along this protocol, we will use this as an example). For larger tissue samples, scale up methanol and chloroform volumes (*see* below).
2. Add 1000–1500 cpm of the T3 tracer in a volume of 10–20 μL in half of the samples. Mix by vortex and leave 5 min to allow for the ^{125}I-T3 to equilibrate with the endogenous T3. Add the same volume and cpms of the T3 tracer in 3 gamma counter-compatible tubes (T3 reference tubes) and keep them apart.
3. Add 1000–1500 cpm of the T4 tracer in a volume of 10 to 20 μL in the other half of the samples. Mix by vortex and leave 5 min to allow for the ^{125}I-T4 to equilibrate with the endogenous T4. Add the same volume and cpms of the T4 tracer in 3 gamma counter-compatible tubes (T4 reference tubes) and keep them apart.
4. Centrifuge at 5000 rpm for 10 min at room temperature.
5. Transfer the supernatant to a 10 mL chloroform-resistant tube.
6. Add 1 volume (500 μL) of methanol to the pellet obtained in **step 4**. Resuspend by vigorously vortexing and centrifuge at 5000 rpm for 10 min at room temperature.
7. Transfer the supernatant to the same chloroform-resistant tube used in **step 5**.
8. Add 2 volumes (1 mL) of chloroform to each tube.
9. Add 0.05% Cl_2Ca to each chloroform-resistant tube. The volume of 0.05% Cl_2Ca added to each sample will be calculated according to the following formula: Cl_2Ca mL = ([V methanol + V chloroform − (g tissue × 0.8)]/4) − (g tissue × 0.8).
10. Mix gently by inversion or gentle vortexing. Then, using a marker indicate the volume level inside the tube for reference.
11. Allow the mixture to settle at 4 °C until two differentiated phases and an interphase are clearly distinguished. This settling process can be left overnight to ensure proper separation.
12. Just before using the T3 and T4 extraction columns prepared in Subheading 3.4, activate them by adding 2 mL of acetate buffer pH 7 to each column. Repeat this step. Place the columns in the rack on top of a collection tray and label each column with the corresponding identification for each sample.
13. When the interphase has clearly settled, collect the upper phase of each tube and place on the corresponding activated T3 and

T4 extraction columns. T3 and T4 will be retained at the anion exchange resin.

14. Add chloroform-methanol-Cl_2Ca (3:49:48) in each chloroform-resistant tube up to the mark indicated in **step 10**.
15. Gently mix as in **step 10** and leave to settle at 4 °C until two differentiated phases and an interphase are clearly distinguished. This process can take approximately 2 h.
16. When the interphase has clearly settled, collect the upper phase of each tube and place on the same activated T3 and T4 extraction columns as in **step 13**. T3 and T4 will be retained at the anion exchange resin. Let the column drain completely before the next step.
17. Add 2 mL of acetate buffer pH 7 to each column. Allow the buffer to flow through the column.
18. Add 2 mL of ethanol to each column. Allow the buffer to flow through the column.
19. Add 2 mL of acetate buffer pH 7 to each column. Allow the buffer to flow through the column.
20. Add 2 mL of acetate buffer pH 4 to each column. Allow the buffer to flow through the column.
21. Add 2 mL of 1% acetic acid to each column in a fume hood. Allow the buffer to flow through the column.
22. Add 2 mL of 35% acetic acid to each column in a fume hood. Allow the buffer to flow through the column. Washes in **steps 17–22** allow for T3 and T4 purification.
23. To initiate the elution of the T3 and T4 retained in the columns, place a gamma counter-compatible tube under each column and add 500 μL of 70% acetic acid to each column to collect the 1st fraction.
24. Place a second gamma counter-compatible tube under each column and collect the 2nd, 3rd, and 4th fractions together by adding 1500 μL of 70% acetic acid to each column. Most of the eluted T3 and T4 should be present in these three fractions.
25. Place a third gamma counter-compatible tube under each column and collect the 5th fraction by adding 500 μL of 70% acetic acid to each column.
26. Place a last gamma counter-compatible tube under each column and collect the 6th fraction by adding 500 μL of 70% acetic acid to each column.
27. Use a gamma counter to measure the T3 and T4 tracer cpms in each tube. Measure also the T3 and T4 reference tubes from **steps** 2 and 3.

28. For each sample, pool all the T3- and T4-containing fractions based on the ^{125}I readings. Calculate the individual percentage of T3 and T4 recovery by comparing each sample with the T3 and T4 reference tubes. For samples containing the T3 tracer, estimate the T4 recovery by determining the mean value from all samples containing the T4 tracer, and vice versa.
29. Totally evaporate the 70% acetic acid in the samples using a SpeedVac vacuum concentrator and resuspend in 310 μL of RIA buffer A by vortexing. For small samples (i.e., less than 150 mg) or those expected to contain a low amount of THs, resuspend in a smaller volume (as 210 μL).
30. Keep the extracted tissue samples at −20 °C until further use.

3.8 Radioimmunoassay for T3 Determination

1. Prepare the standard curve for T3: prepare 5 mL of each solution (0.78 pg/10 μL, 0.39 pg/10 μL, 1.56 pg T3/10 μL, 3.125 pg/10 μL, 6.25 pg/10 μL, 12.5 pg/10 μL, 25 pg/10 μL, and 50 pg/10 μL) using RIA buffer A and the T3 working solution (5 ng/mL).
2. Pipette 10 μL of each T3 standard curve point (0, 0.39, 0.78, 1.56, 3.125, 6.25, 12.5, 25, and 50 pg/10 μL) in RIA tubes in triplicates (*see* **Note 14**).
3. Estimate the anticipated amount of T3 present in the sample. Pipette an appropriate volume of plasma or tissue extract to achieve a T3 content ranging between 1.5 to 12 pg (to fit the linear slope of the curve) in each RIA tube in duplicates. Typically, this corresponds to around 10 μL of plasma extract and between 20 and 50 μL of tissue extract (*see* **Note 15**).
4. Bring the tubes with the standard curve and the target samples to a final volume of 60 μL by adding RIA buffer A.
5. To prepare the blank samples, pipette 110 μL of RIA buffer A in triplicates in RIA tubes.
6. Prepare the T3 antisera in RIA buffer A to pipet 50 μL per tube at the optimal dilution (*see* **Note 3**).
7. Add 50 μL of the T3 antisera to the standard curve and target samples. Do not add T3 antisera to blank samples.
8. Prepare a solution of RIA buffer A at a volume of 100 μL per tube, with a final concentration of 0.1 mg/mL cysteine. Incorporate an appropriate amount of ^{125}I -T3 to ensure a radioactivity count of 8000 cpm per 100 μL.
9. Add 100 μL of the radioactive solution prepared in **step** 7 to each tube. In addition, pipette 100 μL of this solution to 3 new RIA tubes that will serve as standard controls.

10. Mix gently by shaking the racks containing the RIA tubes. Cover them to prevent dust contamination and allow them to incubate overnight at room temperature.
11. The next day prepare a solution containing 30% PEG 6000 and 5% calf serum in SR buffer at a volume of 1.5 mL per tube.
12. Add 1.5 mL of this solution to each tube containing the standard curve samples, the target samples, and the blank samples (do not add this solution to the standard controls). Keep mixing the solution using a magnetic stirrer until it is distributed to all tubes, as it tends to precipitate.
13. Mix each tube well by vortexing.
14. Centrifuge at 2000–2500 rpm at 4 °C for 50 min.
15. Aspirate the supernatant using an aspiration pump coupled to a container for liquid radioactive waste.
16. Count the ^{125}I-T3 cpms in the standard curve samples, the target samples, the blank samples, and the standard controls.

3.9 Radioimmunoassay for T4 Determination

1. Prepare the standard curve for T4: prepare 5 mL of each solution (0 pg/50 μL, 2.5 pg/50 μL, 5 pg/50 μL, 10 pg/50 μL, 20 pg/50 μL, 40 pg/50 μL, 80 pg/50 μL, 160 pg/50 μL and 320 pg/50 μL) using RIA buffer B and the T4 working solution (64 ng/mL).
2. Pipette 50 μL of each T4 standard curve point (0, 2.5, 5, 10, 20, 40, 80, 160 and 320 pg/50 μL) in RIA tubes in triplicates (*see* **Note 14**).
3. Estimate the anticipated amount of T4 present in the sample. Pipette an appropriate volume of plasma or tissue sample to achieve a T4 content ranging between 10 to 40 pg (to fit the linear slope of the curve) in each tube in duplicates. Typically, this corresponds to around 2.5 μL of plasma extract and about 50 μL of tissue extract (*see* **Note 15**).
4. If the volume of extract pipetted is less than 50 μL, adjust the target samples to a final volume of 50 μL by adding RIA buffer B.
5. To prepare the blank samples, pipette 150 μL of RIA buffer B in triplicates in RIA tubes.
6. Prepare the T4 antisera in RIA buffer B to add 100 μL per tube at the optimal dilution (*see* **Note 3**).
7. Add 100 μL of the T4 antisera to the standard curve and target samples. Do not add T4 antisera to blank samples.
8. Prepare a solution of RIA buffer B at a volume of 100 μL per tube, with a concentration of 0.1 mg/mL cysteine. Incorporate an appropriate amount of ^{125}I-T4 to ensure a radioactivity count of 8000 cpm per 100 μL.

9. Add 100 μL of the radioactive solution prepared in **step** 7 to each tube. In addition, pipette 100 μL of this solution to 3 new RIA tubes that will serve as standard controls.
10. Follow **steps 10–15** from Subheading 3.8.
11. Count the ^{125}I-T4 cpms in the standard curve samples, the target samples, the blank samples, and the standard controls.

3.10 Radioimmunoassay Results

Usually, gamma counters include software to adjust the RIA curve, interpolate the unknowns, and obtain the results (pg of T3 or T4) for each sample. Otherwise, plot the standard curve values in a logarithmic scale and adjust for a sigmoidal curve. Interpolate the values for each sample and transform back the logarithmic results into concentrations. The values of the samples should fit the linear slope of the sigmoidal curve for accurate estimation of THs. Apply the corrections for initial volume (for plasma samples) or weight (for tissue samples) pipetted in the RIA, dilution of the sample (if used), and extraction recovery to obtain the final values. Express the results in pg of THs per mL of plasma or pg of THs per mg of tissue.

The limits of detection for these procedures are 2.5 pg T4 and 0.39 pg T3 per tube.

4 Notes

1. Handle Chloramine T appropriately considering its light sensitivity and high hygroscopicity. First, weigh the Chloramine T and Metabisulfite and then calculate the precise volume of 0.05 M PB required to achieve a concentration of 2.5 mg/mL and 20.8 mg/mL, respectively. Despite requiring only 10 μL of Chloramine T and 25 μL of Metabisulfite for each labeling reaction, it is recommended to weigh approximately 2.5 mg of Chloramine T and 20.8 mg of Metabisulfite due to the challenges and variability associated with weighing small quantities. Chloramine T should be protected from light and dissolved in the last moment. Be sure the solution is crystal clear, as turbidity could indicate that the product is not in optimal conditions.
2. BSA prevents non-specific binding to tubes, prevents deiodination of tracers, enhances antigen-antibody interactions, and increases sensitivity of detection. BSA at 22%, pH 7,5 (such as Ortho-Clinical Diagnostics) is used and diluted 1/100, thus avoiding the problems associated to dissolving BSA and adjusting its pH.
3. In the described procedure, the antisera described below have been used. For alternative antibodies, it is recommended to

establish optimal conditions accordingly. Specific antiserum against T4 was raised in rabbits by C. van Hardeveld, D. van der Heide and M. J. Obregon (University Hospital of Leiden, Leiden, The Netherlands) using as antigen BSA conjugated with L-T4 ethyl ester. Specific antiserum against T3 was raised in rabbits by M. J. Obregon (IIB, CSIC-UAM, Madrid) using as antigen L-T3 ethyl ester conjugated with BSA [3, 10]. T3 antiserum is used at a final concentration of 1:300,000 and T4 antiserum at 1:75,000. Cross-reactivity for the T4 RIA was 1.2% for T3 and, for the T3 RIA, cross-reactivity was 1.1% for T4 [10].

4. An appropriate precipitation will yield better RIA results. A quality calf serum is an essential factor in achieving optimum precipitation. Prior to performing the RIA, test several calf serums, as these can exhibit significant variability even within batches. To evaluate calf serums, prepare a solution containing 30% PEG 6000 and 5% of each calf serum in SR buffer. Test each calf serum in triplicates. Add 1.5 mL of this solution per tube, vortex strongly to mix, and centrifuge at 2000–2500 rpm at 4 °C for 50 min. Aspirate the supernatant and assess and compare the presence of precipitate and the absence of turbidity in the supernatant achieved by the different calf serums. Choose the calf serum that better meets these requirements for optimal RIA performance.
5. It is essential that the pH of the labeling reaction remains at 7.5. It has been established (Maria Jesus Obregon, personal communication) that 50 μL of 0.5 M PB is capable of maintaining the pH of the reaction at pH 7.5 regardless of whether 0.5, 1, or 2 mCi and their corresponding substrates are added to the reaction. The amount (mCi) of ^{125}I for the reaction is dependent on the number of samples in which to perform TH determinations.
6. The reaction time is critical. For T3 labeling, note that T2 has a very accessible site (5′) for iodide incorporation and another one less accessible. Optimum reaction time leads to one ^{125}I atom incorporation, generating ^{125}I-T3. Extending the reaction time could lead to the incorporation of two ^{125}I atoms generating doubled-labeled T4. For T4 labeling, the substrate T3 will incorporate a ^{125}I atom generating ^{125}I-T4, however, by extending the reaction time an iodide atom from the T3 substrate could be exchanged by ^{125}I and generate double-labeled T4. During T4 labeling, the T3 substrate could exchange an iodide molecule for ^{125}I without the incorporation of an additional ^{125}I and generate ^{125}I-T3. Hence, the precise control of reaction time is essential to ensure accurate labeling of T3 and T4.

7. As mentioned in note 5, when labeling ^{125}I -T3 from the substrate T2 it is also possible to obtain ^{125}I -T4 and when labeling T4 from T3 it is possible to obtain ^{125}I -T3.
8. Performing the T3 and T4 labeling reactions immediately after the delivery of freshly labeled ^{125}-Iodide will enhance the likelihood of achieving high specific activity labeled products. The specific activity for labeled THs could be low due to iodide contamination (in the substrate, buffers or ^{125}I).
9. When both T3 and T4 determinations will be performed in the same set of samples, half the samples will contain ^{125}I-T3 tracer and the other half ^{125}I -T4 tracer. Using this protocol, we will obtain the individual hormone recovery value following extraction for half of the samples, and the hormone recovery for the other half (which contained a different tracer) will be calculated as the mean percentage of recovery obtained in the first half. If only either T3 or T4 determinations are to be performed, all samples will receive the appropriate labeled tracer and the individual sample recovery value will be obtained for each sample. Another way of obtaining the individual recovery value for each sample is to label either T3 or T4 with I^{131}, which will be only used as a tracer. In this case, it is crucial to perform the extraction process soon after the labeling, as the short half-life of ^{131}I (8 days) results in rapid activity decay.
10. Determine the number of samples required for extraction and aim for a final volume ranging between 10 to 20 μL containing approximately 1000 to 1500 cpm. Bear in mind that about 25% of the volume applied will be retained in the paper and that at least 500 μL ethanol will be necessary to soak and extract the tracers properly.
11. For blood processing, it is important to acknowledge that EDTA-containing plasma could yield aberrant results as it interferes with the RIA. The potential impact of other anticoagulants should be assessed individually. It is also possible to measure THs in serum (after blood coagulation) following the same protocol. Using plasma or serum without the extraction process will induce more variable results due to serum protein interference. TH determinations in blood can be performed following this protocol in human and animal models. To study peripheral TH levels in very young small animals, when the blood volume obtained is insufficient, we recommend measuring liver TH content as this usually reflects the peripheral TH status [11].
12. TH determinations in tissue samples can be performed following this protocol in human and animal models. Tissue employed for THs determinations from juvenile or adult mice should be obtained by perfusing the animals with saline to

remove blood from tissues before their collection (prior to perfusion, blood can be extracted by retroorbital collection and used for THs determinations in plasma). For younger animals, when perfusion is not possible, collect the tissues after euthanizing, freeze them on dry ice, and store at −80 °C. In this case, it should be noted that THs present in the blood could contaminate THs content in the tissue. Note that tissue samples weighing less than 60 mg are likely to yield poor results due to the expected low content of THs. To mitigate this issue, we recommend pooling samples to meet this minimum weight requirement. Accurate tissue sample weight is needed to calculate the amount of $CaCl_2$ for the extraction procedure and to report the final result in pg of THs per mg of tissue.

13. For highly fibrous tissues such as muscle, it is advisable to grind them in a mortar on dry ice before proceeding with homogenization with methanol.
14. Even though the values for the standard curve are pipetted in triplicates, it is expected that the lower values to display high variability. Anticipating this, it is recommended to pipette 5 or 6 samples of the lower values (0, 0.39, 0.78, 1.5 pg/10 μL for T3 RIA and 0, 2.5, 5, 10 pg/50 μL for T4 RIA), count them in a gamma counter and choose those three displaying less variability to generate the standard curve.
15. Prior to analyzing the test samples, it is advisable to run a "sham" RIA using only the standard curve (without samples). This is also recommended when a new standard curve has been prepared and/or a new batch of antibodies are being tested.

Acknowledgments

We are extremely grateful to Maria Jesús Obregon for the careful revision of this manuscript and helpful suggestions. We want to dedicate a special acknowledgment to her mentorship and friendship over all these years. We also want to recognize the dedication of Gabriella Morreale de Escobar and her guidance during our first steps into the world of thyroid hormone research.

References

1. Weeke J, Orskov H (1975) Ultrasensitive radioimmunoassay for direct determination of free triiodothyronine concentration in serum. Scand J Clin Lab Invest 35(3):237–244
2. Obregon MJ, Mallol J, Escobar del Rey F, Morreale de Escobar G (1981) Presence of L-thyroxine and 3,5,3′-triiodo-L-thyronine in tissues from thyroidectomized rats. Endocrinology 109(3):908–913
3. Obregon MJ, Morreale de Escobar G, Escobar del Rey F (1978) Concentrations of triiodo-L-thyronine in the plasma and tissues of normal rats, as determined by radioimmunoassay: comparison with results obtained by an

isotopic equilibrium technique. Endocrinology 103(6):2145–2153

4. Weeke J, Orskov H (1973) Synthesis of 125I monolabelled 3, 5, 3′-triiodothyronine and thyroxine of maximum specific activity for radioimmunoassay. Scand J Clin Lab Invest 32(4):357–360
5. Obregon MJ et al (1984) L-thyroxine and 3,5,3′-triiodo-L-thyronine in rat embryos before onset of fetal thyroid function. Endocrinology 114(1):305–307. https://doi.org/10.1210/endo-114-1-305
6. Obregon MJ et al (1991) Outer ring iodothyronine deiodinases and thyroid hormone economy: responses to iodine deficiency in the rat fetus and neonate. Endocrinology 129(5):2663–2673. https://doi.org/10.1210/endo-129-5-2663
7. de Escobar GM et al (2008) The changing role of maternal thyroid hormone in fetal brain development. Semin Perinatol 32(6):380–386. https://doi.org/10.1053/j.semperi.2008.09.002
8. Aranda A et al (1981) Estudio piloto para la deteccion precoz de hipotiroidismo congénito en España. Endocrinologia 28:85–90
9. Gutleb AC, Cambier S, Serchi T (2016) Impact of endocrine disruptors on the thyroid hormone system. Horm Res Paediatr 86(4):271–278. https://doi.org/10.1159/000443501
10. Ruiz de Oña C et al (1991) Thyroid hormones and 5′-deiodinase in the rat fetus late in gestation: effects of maternal hypothyroidism. Endocrinology 128(1):422–432
11. Bárez-López S, Obregon MJ, Bernal J, Guadaño-Ferraz A (2018) Thyroid hormone economy in the perinatal mouse brain: implications for cerebral cortex development. Cereb Cortex 28(5):1783–1793. https://doi.org/10.1093/cercor/bhx088

Chapter 4

Generation of a Mouse Model for the Study of Thyroid Hormones Regulatory Effect on the Immune System

Cristina Saiz-Ladera

Abstract

The generation of hypothyroid and hyperthyroid mouse models is one of the approaches used to investigate the complex interplay between thyroid hormones and the immune system. We present a detailed protocol describing how to induce endotoxic shock by lipopolysaccharide (LPS) administration, and how to investigate the role of immune populations, specifically macrophages, responding to endotoxemia.

This book chapter provides the use of different molecular techniques, such as Western Blotting, Immunohistochemistry, q-PCR, Luciferase assays, or ChIP assays, with which researchers can gain valuable insights into the immune system's interaction with hormonal signaling pathways, for instance, examining the effect of thyroid hormones on signaling of STAT3, NF-κB, and ERK in response to LPS, and inflammatory mediators, such as interleukin-6 (IL-6) or tumor necrosis factor-alpha (TNFα) within these cells. The signaling pathways involved and the exploration of the relationship between thyroid hormones and the immune system can be analyzed using several molecular biology technologies in order to clarify their interplay in various disease states.

Key words Endotoxemia, Lipopolysaccharide (LPS), Thyroid hormone receptors (TRs), Immune response, Hyperthyroidism, Hypothyroidism, Mouse model

1 Introduction

Mouse lacking TRα 1 and TRβ, the main thyroid hormone binding isoforms [1], is an animal model used to analyze whether TRs could modulate the activity of signaling pathways involved in immune responses, such as the endotoxic shock, the most lethal form of sepsis caused by the main membrane component of Gram-negative bacteria: the lipopolysaccharide (LPS) [2], and where an excessive inflammatory response to an infection occurs [3].

In vivo experimental models are tools that serve as a living organism in which to reproduce an induced or in an inherent natural acquisition way spontaneous pathologic processes that could resemble the same phenomenon occurring in humans [4]. Different experimental methods have been used to induce

Constanza Contreras-Jurado (ed.), *Thyroid Hormones: Methods and Protocols*, Methods in Molecular Biology, vol. 2876, https://doi.org/10.1007/978-1-0716-4252-8_4,

hypo- and hyperthyroidism in animal models. The most common methods that have been used to induce hypothyroidism in animal models are: (1) Surgical model—surgical removal of thyroid gland; (2) Radioactive model—removal of thyroid gland by means of iodine radioactive isotope (131I); (3) Dietetic iodine restriction model; (4) Substance administration model—antithyroid drug administration such as methylimidazole (MMI), propylthiouracyl (PTU), or $KClO_4$; 5) Genetic model—genome manipulation [4]. The most common methods that have been used to induce hyperthyroidism in animal models are: (1) Dietetic iodine model—exogenous thyroxine of thyroxine, T4, and triiodothyronine, T3, in drinking water, besides intragastric incubation of T4; (2) Genetic model—genome manipulation- [5, 6].

The actions of the thyroid hormones (L-thyroxine, T4, and 3,3′,5-triiodo-L-thyronine, T3) are mediated by binding to nuclear receptors (TRα and TRβ). These receptors act as ligand-dependent transcription factors by binding, generally as heterodimers with the retinoid X receptor (RXR), to thyroid hormone response elements in target genes or by modulating the activity of other transcription factors or signaling pathways (1). The liver plays a critical role in innate and adaptive immunity, and in response to infection and inflammation, it synthetizes acute-phase proteins (APPs), which are key components of the immune response to infection [7–9]. Innate immune responses triggered by LPS are mediated by Toll-like 4 receptors and involve the coordinated production of a large variety of inflammatory mediators, particularly interleukin 6 (IL-6) and tumor necrosis factor α (TNFα) [10, 11]. Activation of signal transducer and activator of transcription 3 (STAT3) and nuclear factor kappa-light-chain-enhancer of activated B cells (NF-κ B) by these cytokines plays a key role in the liver response to inflammation [9, 11], controlling the expression of a large number of genes [12].

To analyze whether TRs could modulate the activity of signaling pathways involved in immune responses, immunoblotting techniques, such as western blotting, immunohistochemistry, or mouse cytokine array panel that quantify cytokines and chemokines delivered in this process, are performed to measure the levels of total and phosphorylated main proteins involved in the liver response to inflammation [13]. There is an increasing evidence that the thyroid hormones could modulate immune responses [14–16], and the molecular mechanisms involved in the cross-talk between thyroid hormone and immune function can also be analyzed by immunoblotting and immunochemistry techniques, which provide information about the cytokines [17] that change thyroidal T4 and T3 synthesis, that the expression of deiodinases thyroid hormone metabolism is altered [18–21] or TRβ and RXR expression is reduced [22, 23], processes that can be demonstrated with using qRT-PCR, luciferase assays or ChIP arrays methodology which clarify this links. The use of all these research tools can help us to verify that TRs can regulate inflammatory responses and these receptors control hepatic cytokine production and signaling during sepsis in vivo, in addition, that the thyroid hormone directly antagonizes IL-6 signaling in cultured hepatocarcinoma cells, and for instance, immune populations such as macrophages regulate the transcriptional activity of STAT [2].

2 Materials

2.1 Animal Models

All experiments can be performed in the following in vivo animal models:

1. Adult TRα 1−/−/TRβ −/− double knockout (KO) mice and in wild-type TRα 1+/+/TRβ +/+ animals in a hybrid genetic background of 129/OLa + 129/Sv + BALB/c + C57BL/6,
2. 8 weeks old wild-type male C57BL/6 J mice.

All animal work has to be carried out in compliance with the European Community Law (86/609/EEC) and the Spanish law (R.D. 1201/2005), with approval of the Ethics Committee of the Consejo Superior de Investigaciones Científicas and were bred at the Instituto de Investigaciones Biomédicas.

2.2 Cell Culture

All experiments can be performed using in vitro conditions:

1. Hep-3B, RAW264.7 and L929 fibroblasts cell lines.
2. MW6, 100 and 150 mM dishes.
3. AG1-X8 resin.
4. AG1-X8 resin treated FCS (Fetal Calf Serum): treated FCS.
5. DMEM medium with 10% treated FCS.
6. RPMI medium containing 10% treated FCS.

2.3 Reagents, Solutions, and Preservation Media

Prepare all solutions using ultrapure water prepared by purifying deionized water.

Prepare and store all reagents at room temperature (unless indicated otherwise). Rigorously, follow all waste disposal regulations when disposing waste materials.

1. Hypotonic buffer: prepare solution A, solution B and mix in a solution C. Solution A: Add about 100 mL water to a 500 mL graduated cylinder or a glass beaker (*see* **Note 1**). Weigh 4.95 g NH4Cl and transfer to the cylinder. Add water to a volume of 400 mL. Make up to 500 mL with water. Store at 4 °C. Solution B: Add about 10 mL water to a 100 mL graduated cylinder or a glass beaker (*see* **Note 1**). Weigh 2.06 g of Tris Base and transfer to the cylinder. Add water to a volume of

90 mL. Make up to 100 mL with water. Store at 4 °C. Mix and adjust pH with HCl to pH 7.65 (*see* **Note 2**). Solution C: mix 90 mL of solution A + 10 mL of solution B, mix and adjust with HCl to pH 7.2 (*see* **Note 2**). Store at 4 °C.

2. Antithyroidal drug: 0,15% of propylthiouracil (E15551-04, Sniff) in 1% $KClO_4$ (10 g/liter) (460494, Merck) solution: Add about 300 mL water to a 1 L graduated cylinder or a glass beaker (*see* **Note 1**). Weigh 150 mg of propylthiouracil and 10 g of $KClO_4$ and transfer to the cylinder. Add water to a volume of 900 mL. Make up to 1 L with water. Store at 4 °C.
3. T4 (25 ng/g of mice; IRMM468, Sigma-Aldrich) and **T3** (95 ng/g of mice; T-2877, Sigma-Aldrich).
4. LPS (Lipopolysaccharide) solution: at concentration range of 1-20 mg/kg BW from *Escherichia coli* serotype 026:B6, according to Sigma–Aldrich, San Luis, MO.
5. Control diet (E15552–24, Sniff): supplemented with potassium iodide: supplemented mice diet with 1.15 mg/kg. Weigh the adult mice (g) and calculate the dosage to be given in 1.15 mg/kg.
6. CO_2 atmosphere compartment: for the sacrifice of experimental animals.
7. 4% buffered formalin solution in PBS 1× (Phosphate-Buffered Saline Ph 7.4).
8. 70% ethanol in ultrapure water.
9. Paraffin.
10. H&E: hematoxylin and eosin dyes.
11. *RNA later* solution.
12. Freezer at −80 °C.
13. Liquid nitrogen canister.

2.4 Protein Extraction

1. RIPA buffer: 50 mM Hepes pH 7.5, 150 mM NaCl, 10% Glycerol, 1 mM EGTA, 1% Triton-X100, 1% Sodium Deoxycholate, 0,1% SDS.
2. Triple-detergent lysis buffer: 50 mM Tris-HCl pH 8.0, 150 mM NaCl, 0.1% SDS, 1% NP-40, 0.5% sodium deoxycholate.
3. Protease (aprotinin and leupeptin 5ug/mL, PMSF 1 mM, DTT 2,5 Mm) and Phosphatase inhibitors (Na3VO4 1 Mm, NaF 20 Mm and NaPPi 1 mM), in cocktail tablets (according to *Roche*, Basel, Switzerland protocol).
4. *Bradford* colorimetric analysis (according to *BioRad* protocol).
5. Standard curve of BSA (Bovine Serum Albumin) solution (1,25ug/ul- 20 ug/ul in water).

6. Liquid nitrogen canister.
7. Bath at 37 °C.
8. Centrifuga.

2.5 Western Blotting

1. 8% or 12% sodium dodecyl sulfate-polyacrylamide-SDS precast gel (in accordance with the supplier's instructions).
2. Nitrocellulose membranes (*Amershan Pharmacia Biotech*).
3. Molecular weight marker (10 kDa–250 kDa, *BioRad*).
4. Semi-wet chamber and source of electrophoresis.
5. Western blot transfer buffer: 0.025 M Tris–HCl, 0.192 M glycine, 20% methanol (*see* **Note 3**).
6. Tris-buffered saline (TBS; 10×): 1.5 M NaCl, 0.1 M Tris–HCl, pH 7.4.
7. TBS containing 0.05% Tween-20 (TBST).
8. Blocking solution: 5% milk or BSA in TBST. Add 100 mL of 10× TBS to a 1 L graduated cylinder and make it to about 800 mL with water. To prepare the blocking solution TBST with 5% milk, or alternatively, with 5% BSA: transfer 50 g skim milk or BSA powder into the cylinder and stir until dissolved. Make to 1 L with water. Separate 500 mL as the blocking solution. To the remaining 500 mL add 250 μL of Tween-20 (cut end of blue tip to aspirate Tween-20 easily), dissolve, and use it as the diluent. Store at 4 °C.
9. Antibodies diluent solution: 5% BSA in TBST. Store at 4 °C.
10. Laemmli sample buffer (according to the supplier's instructions).
11. Transfer buffer: 33 mM Tris-Base, 194 mM Glycine and 20% of methanol.
12. Antibodies and dilutions: 1:1000 Anti-P-STAT3Tyr705 Rabbit (Cell Signaling 9145 s); 1:1000 Anti-STAT3 Rabbit (Cell Signaling C7907); 1:1000 Anti- P-ERK Mouse (Santa Cruz, sc-7383); 1:1000 Anti-ERK2 Rabbit (Santa Cruz sc-154); 1: 500 Anti-P-p65 Rabbit (Cell Signaling 3033); 1:1000 Anti-p65 Mouse (Cell Signaling 6956P); 1:500 Anti-IKBα Rabbit (Santa Cruz sc-371); 1:500 Anti-IKBβ Rabbit (Santa Cruz sc-945).
13. Luminescence detection system (*Super Signal West Pico Chemiluminescent Substrate, Pierce*).
14. Luminescence capture system using X-ray films or digital images.
15. *ImageJ* app tool.
16. Stripping solution: 62.5 mM Tris HCl pH 6,7, 2% SDS, 100 Mm β-Mercaptoethanol.

2.6 Immunohistochemistry

1. Sodium citrate buffer: 10 mM, pH 6,0.
2. 3% oxygenated water (H2O2).
3. Antibodies and dilutions: F4/80 marker at 1:100.
4. PBS 1x (Phosphate-Buffered Saline Ph 7.4).
5. Blocker solution: 5% BSA (Bovine Serum Albumin) in PBS 1x.
6. Diluent solution: 1% BSA in PBS 1x.
7. ABC *kit*: Avidin/streptavidin-peroxidase system (according to the supplier's instructions from *Vectastain*).
8. DAB, Diaminobencidine, Peroxidase Substrate *kit* (according to the supplier's instructions, *Vector Laboratories*).
9. DePeX (*Serva*), mounting media.
10. Microscope with digital camera system (f.i. *Leica DC200* mounted on an *Olympus DMLB*).

2.7 Cytokines and Glucose Quantification and Gene Expression Analysis

1. R&D Systems Mouse Cytokine Array, Panel A (following the manufacturer specifications, *Catalog #ARY006, Minneapolis, MN*).
2. X-ray films or image capture system.
3. *ImageJ* app tool.
4. *Reflotron test* (Roche).
5. *miRNAeasy Mini Kit*, Qiagen manufacture.
6. *Rnase-Free Dnase Set*, Qiagen manufacture.
7. Agilent Bioanalyzer.
8. *Qiagen Omniscript RT Kit* retrotranscription system.
9. RNAase-free water.
10. Specific primers (500 mM).
11. cDNA (1ug),
12. *Power SYBR Green PCR Master Mix* from *Applied Biosystems.*
13. Applied Biosystems 7500 Fast Real-time PCR System.

2.8 Transient Transfections and Luciferase Assays

1. Luciferase reporter plasmids with consensus NF-kB and STAT3 binding sites.
2. pRL-TK-Renilla (*Promega*, San Luis, CA) normalizer control.
3. Lipofectamine 2000 (*Invitrogen*, Carlsbad, CA).
4. AG1-X8 resin (*Bio-Rad*, Hercules, CA).
5. 10 ng/mL of IL6.
6. 10 ng/mL of TNFα.
7. Dual Luciferase Assay System (*Promega*).

2.9 ChIP Assays

1. 17–295 *Upstate* fixing and lysing kit.
2. STAT3 antibody (sc-482(c-20)x).
3. Acetylated histone H4 antibody (Upstate 06–866).
4. Normal rabbit IgG (sc-2027) control.
5. Bioruptor UCD-200TM (*Diagenode*, Liège, Belgium).
6. C-Reactive Protein (CRP) oligonucleotides: 5′ - GAAA TAATTTTGCTTCCCCTCTTCCC-3′ (forward) and 5′ -TC CTAGATCTCTTGCCTTAGAGCTACCTCC-3′ (reverse).
7. Promoter of the *Hepcidine* gene oligonucleotides: 5′ -CCCCACCCCCTGAACACA-3′ (forward) and 5′ -ACCGAGTGACAGTCGCTTTT-3′(reverse).

3 Methods

Carry out all procedures at room temperature unless otherwise specified.

3.1 Generation of Hypothyroid, Euthyroid, and Hyperthyroid Mice

All studies and experiments are targeted and performed in livers from adult TRα 1−/−/TRβ −/− double knockout (KO) mice and in wild-type TRα 1+/+/TRβ +/+ animals, genetically modified organisms (GMOs) in a hybrid genetic background of 129/OLa + 129/Sv + BALB/c + C57BL/6.

1. To induce hypothyroidism, 4- to 5-week-old male mice are fed with an iodine-deficient diet containing 0.15% of the antithyroidal drug propylthiouracil with 1% $KClO_4$ in the drinking water for 4 weeks before the experiment.
2. Control euthyroid animals are fed with a control diet, the same diet without propylthiouracil but supplemented with 1.15 mg/kg potassium iodide.
3. Eight weeks old wild-type male C57BL/6 J mice bred are made hyperthyroid by adding 25 ng/g of mice T4 and 95 ng/g of mice T3 in the drinking water, from 14 days before infection until the end of the experiment (*see* **Note 4**).
4. Drinking water is changed every day. Unless otherwise indicated, mice are maintained with the chosen diet until the end of the experiment.

3.2 Cell Culture

1. Before the experiments, cells are transferred to medium containing FCS (Fetal calf serum) depleted of thyroid hormones by treatment with AG1-X8 resin.
2. Hep-3B cells grow in DMEM media and RAW264.7 cells in RPMI media, both with 10% FCS: add 50 mL of FCS in 450 mL of DMEM and RPMI.

3. Bone marrow macrophages are obtained from the tibia and femur of 8 weeks old C57-BL6 mice, plated and grown as previously described in DMEM and 30% medium conditioned by L929 fibroblasts.
4. After 6 days the medium is changed to DMEM resin treated FCS and start treatments.

3.3 Generation of Endotoxemia in Animal Models

Animals are injected i.p (*see* **Note 5**) with the LPS (at concentrations varying from 1 to 20 mg/kg BW) from *Escherichia coli serotype* 026: B6 and they are sacrificed at different times.

3.4 Sample Collection from Animal Models

1. Animals are sacrificed in a CO2 atmosphere as required for the experiments or show signs of deterioration.
2. Necropsies: serum and liver samples are collected in different preservation media, depending on the type of processing and subsequent experimental analysis.
3. Liver samples are collected and fixed in 4% buffered formalin solution in PBS 1x or in 70% ethanol, and embedded in paraffin to histopathological analysis with hematoxylin and eosin (H&E) staining and to perform immunohistochemistry analysis.
4. Liver samples are collected in *RNA later* solution for RNA extraction and analysis, and stored deep freezer at −80 °C, for molecular expression analysis.
5. Serum and liver samples for biochemical analysis are frozen in liquid nitrogen or at −80 °C until use.

3.5 Immunohistochemistry Analysis of Macrophages Immune Population

1. Deparaffinize and rehydrate 4 μm sections of liver tissue. Then, deparaffinized-rehydrated sections are stained with H&E using standard procedures. Images are obtained using a high-resolution digital camera system mounted on a microscope or equivalent.
2. To antigen retrieval, incubate sections with 10 mM; pH 6,0 Sodium citrate buffer to unmask epitopes for 30–45 min.
3. Inhibit endogenous peroxidase activity with 3% $H2O_2$ for 10 min.
4. Wash three times with PBS 1×, 5 min.
5. Incubate samples with 5% of BSA to block non-specific binding.
6. Immunohistochemistry standard procedure of the F4/80 macrophage marker is performed on the 4 μM deparaffinized-rehydrated sections with 1:100 antibody dilution in 1% BSA in PBS 1×. (All needed antibodies are administered in 1% BSA in PBS 1x solution).

7. Wash three times with PBS 1×, 5 min.
8. Incubate samples overnight with the F4/80 primary marker antibody at 4 °C (*see* **Note 6**).
9. Wash three times with PBS 1×, 5 min.
10. Use ABC *kit* to amplify the signal according to the supplier's instructions.
11. Wash three times with PBS 1×, 5 min.
12. Reveal the slides with DAB *kit* as chromogenic substrate, according to the supplier's instructions.
13. Wash three times with PBS 1×, 5 min.
14. Counterstain samples with hematoxylin, wash three times with PBS 1× for 5 min and mounted with DePeX according to the supplier's instructions.

3.6 Protein Extraction

1. Supplement with protease inhibitors (aprotinin, leupeptin, PMSF, DTT) and phosphatase inhibitors (Na3VO4, NaF, NaPPi), triple-detergent lysis buffer, and RIPA buffer.
2. Immerge samples in a triple-detergent lysis buffer, in case of tissue, or RIPA buffer, in case of cells. Homogenize actively liver tissue with mortar or pipette in case of cells, and passively with 3 freeze-thaw cycles using liquid nitrogen and bath at 37 °C.
3. Centrifuge the extract proteins in triple-detergent lysis or RIPA buffer at 12.000 rpm at 4 °C for 10 min.
4. Collect the supernatants and transfer to a cold tube.
5. Determine protein concentration by *Bradford* colorimetric analysis using a standard curve ranging from 1,25ug/ul-20 ug/ul of BSA and following the manufacturer's recommendations.

3.7 Western Blotting

1. Quantify 50ug of protein per sample.
2. Mix protein lysates with Laemmli simple buffer and boil.
3. Load samples and molecular weight marker onto 8% or 12% sodium dodecyl sulfate-polyacrylamide gel electrophoresis and apply 120 V to protein separation by electrophoresis for 1 h.
6. Transfer the contents of the gels to nitrocellulose membranes by semi-wet chamber electrophoresis using a transfer buffer at 90 V for 90 min at 4 °C and following the manufacturer's recommendations of *Bio-Rad.*
4. Wash three times with TBST 1x, 5 min.
5. Block nonspecific bindings with 5% BSA TBST for 1 h at RT.
6. Wash three times with TBST 1x for 5 min.

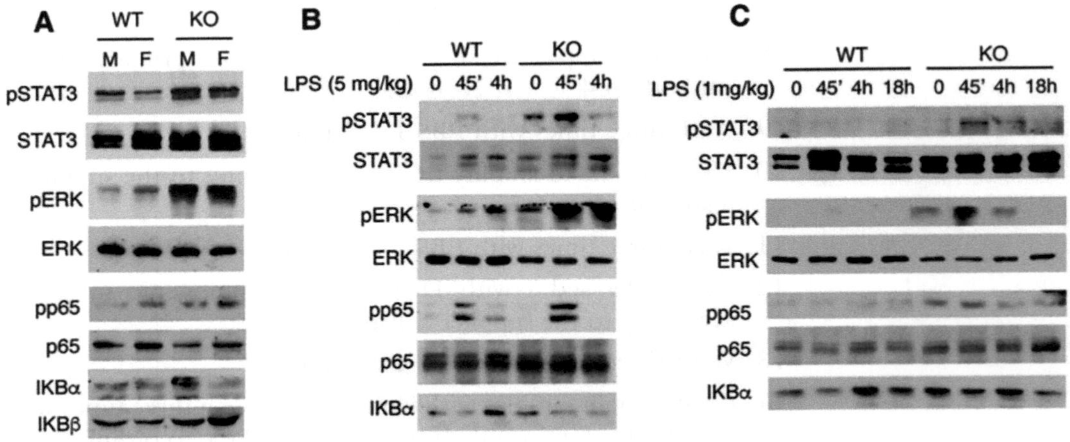

Fig. 1 Example of exposition of TRs deletion increases hepatic STAT3 and ERK activation by Western Blot. (**a**) Western blots with liver extracts of WT and TR KO mice and the indicated antibodies. Both male (M) and female (F) mice were used. (**b**) Protein levels assayed in male mice treated with vehicle or 5 mg/kg BW of LPS for the indicated time periods. (**c**) Similar experiment in WT and KO mice treated with 1 mg/kg BW LPS. Each band represents pooled samples from 3 different animals. (With permission and reproduced from ref. [2])

17. Incubate nitrocellulose membranes with the following primary antibodies in 1% BSA TBST, in this order and sequentially: 1:1000 Anti-P-STAT3Tyr705 Rabbit; 1:1000 Anti-P-ERK Mouse; 1:500 Anti-P-p65 Rabbit; 1:1000 Anti-STAT3 Rabbit; 1:1000 Anti-ERK2 Rabbit; 1:1000 Anti-p65 Mouse; 1:500 Anti-IKBα Rabbit; 1:500 Anti-IKBβ Rabbit, and shaking for 12 h at 4 °C.

7. Wash three times with TBST 1x, 5 min.

8. Detect the primary antibodies signal by pre-oxidase-bound secondary antibodies for 1 h at RT.

9. Visualize the signal with a luminescence detection system based on a specific peroxidase binding using *Super Signal West Pico Chemiluminescent Substrate (Pierce)* and X-Ray films or other image capture system (Fig. 1).

10. Quantify the chemiluminescence signal scanning at different exposures and analyze by *ImageJ* app tool.

11. In the case of using the same membrane for the detection of several proteins consecutively, remove the previously bound antibodies by treating the membrane with a stripping solution at 55 °C for 30 min under agitation.

3.8 Cytokine and Chemokine Quantification

1. Use 100 μl of serum or 400 μg of whole-cell protein extracts obtained from livers from 3–4 animals per group and following the manufacturer specification of *R&D Systems Mouse Cytokine Array, Panel A* to detect simultaneously the levels of 40 different cytokines and chemokines released into the extracellular milieu.

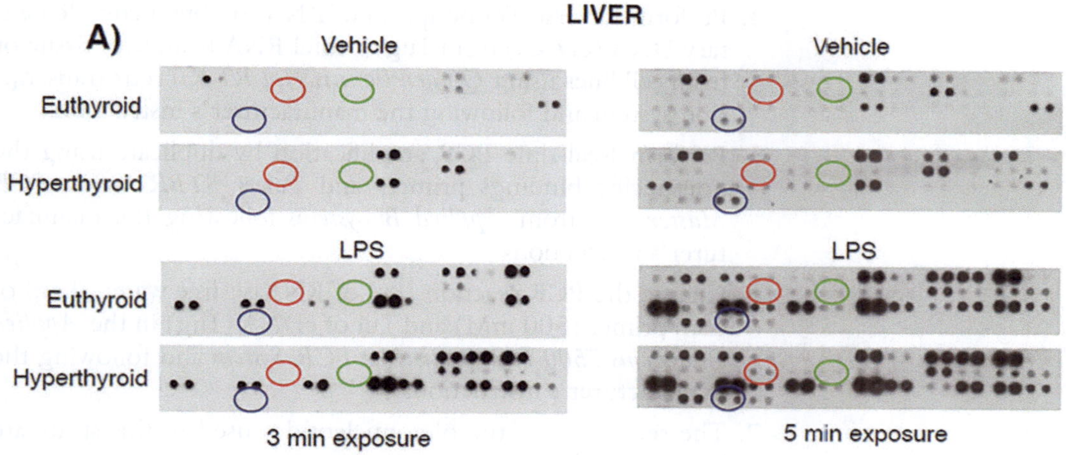

Fig. 2 Example of arrays exposition of cytokines and chemokines
Hepatic and circulating cytokines and chemokines in hyperthyroid mice. A) Liver extracts from euthyroid and hyperthyroid mice (pools from 4–6 animals) were used to detect cytokines and chemokines with Mouse Cytokine Arrays. Mice were treated with vehicle or with LPS (5 mg/kg) for 5 h. Two different exposures of the arrays (3 min and 5 min) are shown. The blue, red, and green circles show location of the spots corresponding to TNFα, IL-6, and IL-10, respectively. Note that IL-6 is only clearly detectable at high exposures in hyperthyroid LPS-treated mice corresponding with the high IL-6 mRNA levels in the livers of these animals. (With permission and reproduced from ref. (2)

2. Detect chemiluminescence signals by the previously described protocol in Subheading 3.7 (Fig. 2).
3. Quantify the chemiluminescence signal scanning at different exposures and analyze by *ImageJ* app tool.

3.9 Aspartate Aminotransferase (ATS) Activity

1. Measure serum AST activity in 32 μl of a 1:3 dilution of sera with the *Reflotron test* (Roche) following the manufacturer's instructions.

3.10 Real-Time Quantitative RT-PCR (qRT-PCR) to Gene Expression Analysis

1. Carry out RNA extraction from liver tissue or cell lines after mechanical disaggregation and perform total RNA purification using a tissue- and/or cell-specific *miRNAeasy Mini Kit, Qiagen*, according to the manufacturer's instructions.
2. Remove genomic DNA from the samples by DNAase treatment using *Rnase-Free Dnase Set, Qiagen* according to the manufacturer's instructions.
3. Check the integrity of the extracted RNA using an *Agilent Bioanalyzer* system following the quality criteria based on values from 1 of the 28S/18S ratio or values from 8 of the RNA integrity number (RIN) as a measure of the degree of degradation of the sample (Schroeder, Mueller et al., 2006).

4. Perform Reverse Transcription of RNA to obtain complementary DNA (cDNA) from 1ug of total RNA from liver tissue or from cell lines using *Qiagen Omniscript RT Kit* retrotranscription system and following the manufacturer's instructions.
5. Perform Real-time PCR amplification by duplicate using the nonspecific bindings primers and *Power SYBR Green PCR Master Mix* from *Applied Biosystems* following the manufacturer's instructions.
6. Charge the PCR reaction (3ul of RNAase-free water, 0.5ul of each primer (500 mM) and 1ul of cDNA (1ug) in the *Applied Biosystems 7500 Fast Real-time PCR System* and following the manufacturer's instructions.
7. The sequences of the oligonucleotides used in this study are listed in Table 1.
8. Program the PCR reaction conditions in *Applied Biosystems 7500 Fast Real-time PCR System*: 1 cycle 50 °C 20 min; 1 cycle 90 °C 10 min; 40 cycles 95 °C 15 min; 1 cycle 60 °C 1 min; 1 cycle 95 °C 15 seconds; 1 cycle 60 °C 1 min; 1 cycle 95 °C 30 seconds; 1 cycle 60 °C 15 seconds.
9. Detect fluorescence signal at the end of each amplification step, and *Applied Biosystems 7500 Fast Real-time PCR System* provides several representation plots and graphs. Perform a *Melting* test to verify the specificity of the absence of dimers in the primers. Calculate from the obtained *Ct* (*Cycle threshold*) values by the apparatus, the efficiency ratio for each primer pair of each gene with the normalizing gene.

3.11 Transient Transfections and Luciferase Assays

1. Perfom all experiments in triplicate and repeated three times.
2. Plate Hep-3B cells in DMEM with 10% FCS in 24-well plates.
3. After 24 h, transfect cells using Lipofectamine 2000 in serum-free medium with 300 ng of luciferase reporter plasmids containing consensus NF-kB and STAT3 binding sites and 50 ng of pRL-TK-Renilla as a normalizer control, following the manufacturer's instructions.
4. After 6 h, transfer cells to supplemented medium with 0.5% thyroid hormone-stripped serum by treatment with AG1-X8 resin.
5. Incubate cells with 5 nM T3 for 36 h in the presence and absence of 10 ng/mL of IL-6 or TNFα for the last 5 h.
6. Determine luciferase activity with the *Dual Luciferase Assay System* following the manufacturer's instructions.
7. Express data as mean ± s.d. and as fold induction over the values obtained in the untreated cells.

Table 1
Example of list of primers used for real-time quantitative RT-PCR (qRT-PCR). (With permission and reproduced from ref. [2])

	Amplification primers	
	Sequence (5′-3′)	
GENE	Forward	Reverse
Mouse IL6	GCTACCAAACTGGATATAA TCAGGA	CCAGGTAGCTATGGTAC TCCAGA
Mouse IL10	TCCTCCTCCAGCTCTTACCTC	TGGCTTTCCCTAGGACTCTCT
Mouse TNFα	CGATCACCCCGAAGTTCAGTA	GGTGCCTATGTCTCAGCCTCTT
Mouse Socs3	ATTTTCGCTTCGGGACTAGC	AACTTGCTGTGGGTGACCAT
Mouse GAPDH	TTCACACCATGGAGAAGGC	CCCTTTTGGCTCCACCCT
Mouse Dio1	GTTGAACTTTGGCAGTTGCAC	GGCTGTGGAGGCAAAGTCATC
Human Gp130	TCAACTTGGAGCCAGATTCC	CCCACTTGCTTCTTCACTCC
Human I16-r	TTGTTTGTGAGTGGGGTCCT	TGGGACTCCTGGGAATACTG
Human CRP	CCCTGAACTTTCAGCCGAATACA	CGTCCTGCTGCCAGTGATACA
Human Haptoglobin	TTGCAGTGGACTCAGGCAAT	CAGCCGTCATCTGCTTCACAT
Human Hepcidin	CCCCACCCCCTGAACACA	ACCGAGTGACAGTCGCTTTT
Human SAA1	CTGCAGAAGTGATCAGCG	ATTGTGTACCCTCTCCCC
Human socs3	ATCCTGGTGACATGCTCCTC	GGCACCAGGTAGACTTTGGA
Human β-fibrinogen	AGCAGCTGCCACTCAAAAGA	GAGGAGGTCTGGGAAACAGC

3.12 ChIP Assays

1. Plate Hep-3B cells within DMEM and 10% of FCS in 150 mm dishes.
2. After 24 h, wash and treat with 5 nM T3 for 36 h and/or 10 ng/mL IL6 for 2 h in medium containing 0.5% thyroid hormone-depleted serum.
3. Fix and lyse the cells following the specifications of 17–295 *Upstate kit.*
4. Sonicate in a *Bioruptor UCD-200TM* following the manufacture's specifications.

5. Prepare each immunoprecipitation assay with 3×10^6 cells, 5ug of STAT3 antibody, 1ug of acetylated histone H4 antibody or 3 ug of normal rabbit IgG to purify and precipitate DNA.
6. Amplify the promoter region of the C-Reactive Protein (*CRP*) promoter with the oligonucleotides 5′-GAAA-TAATTTTGCTTCCCCTCTTCCC-3′(forward) and 5′-TCCTAGATCTCTTGCCTTAGAGCTACCTCC-3′ (reverse), and the promoter of the *Hepcidine* gene with the oligonucleotides 5′ -CCCCACCCCCTGAACACA-3′ (forward) and 5′ -ACCGAGTGACAGTCGCTTTT-3′(reverse).
7. Measure the luciferase activity by *Berthold* luminometer following the manufacture's specifications.
8. Normalize and present the results as a fraction of the inputs after subtracting the values obtained with the control IgG, which is always below 1% of the input chromatin.

4 Notes

1. Having water at the bottom of the cylinder helps to dissolve the Tris relatively easily, following the magnetic stir bar to go to work immediately. If using a glass beaker, the Tris can be dissolved faster if the water is warmed to about 37 °C. However, the downside is that care should be taken to bring the solution to room temperature before adjusting pH.
2. Concentrated HCl (12 N) can be used at first to narrow the gap from the starting pH to the required pH. From then it would be better to use a series of HCl (e.g., 6 N and 1 N) with lower ionic strengths to avoid a sudden drop in pH below the required pH.
3. Dilute 100 mL of 10× native buffer TBS to 800 mL with water and add 200 mL of methanol. Avoid adding methanol directly to the 10× buffer, since it precipitates its ingredients. Even in such a scenario the precipitate can be redissolved by the addition of 800 mL water.
4. Alternative hyperthyroid mice induction: exogenous thyroxine (T4) in drinking water (0.002% w/v) beside intragastric incubation of 50–200 T4 _g/kg body weight (b. wt.) [5].
5. I.p. injection: Intraperitoneal Via: We tilt the animal approximately 45° downward, so that the viscera descend, and we proceed to prick the lower and left area of the abdomen, as there are no vital organs here. Use an insulin syringe.
6. Additional alternative: incubate the primary antibody at room temperature (RT) for 1 h. The trade-off is that this method reduces protocol times but may introduce more non-specific

binding. Primary antibodies need only be used when the primary antibody is directly conjugated to the signaling molecule. Otherwise, an additional incubation with a secondary antibody is required for one hour at RT.

References

1. Pascual A, Aranda A (2013 Jul) Thyroid hormone receptors, cell growth and differentiation. Biochim Biophys Acta 1830(7): 3908–3916
2. Contreras-Jurado C, Alonso-Merino E, Saiz-Ladera C et al (2016) The thyroid hormone receptors inhibit hepatic Interleukin-6 signaling during Endotoxemia. Sci Rep 6:30990
3. Gotts JE, Matthay MA (2016) Sepsis: pathophysiology and clinical management. BMJ 353: i1585
4. Argumedo GS, Sanz CR, Olguín HJ (2012) Experimental models of developmental hypothyroidism. Horm Metab Res = Horm und Stoffwechselforsch = Horm Metab 44(2): 79–85
5. Ahmed OM, Ahmed RG, El-Gareib AW et al (2012) E ffects of experimentally induced maternal hypothyroidism and hyperthyroidism on the development of rat offspring: II-the developmental pattern of neurons in relation to oxidative stress and antioxidant defense system. Int J Dev Neurosci Off J Int Soc Dev Neurosci 30(6):517–537
6. Aranda A, Martínez-Iglesias O, Ruiz-Llorente L et al (2009) Thyroid receptor: roles in cancer. Trends Endocrinol Metab 20(7):318–324
7. Bode JG, Albrecht U, Häussinger D et al (2012) Hepatic acute phase proteins–regulation by IL-6- and IL-1-type cytokines involving STAT3 and its crosstalk with NF-κB-dependent signaling. Eur J Cell Biol 91(6–7): 496–505
8. Zhang YY, Ning BT (2021) Signaling pathways and intervention therapies in sepsis. Signal Transduct Target Ther 6(1):407
9. Medzhitov R (2007) Recognition of microorganisms and activation of the immune response. Nature 449(7164):819–826
10. Pålsson-McDermott EM, O'Neill LAJ (2004) Signal transduction by the lipopolysaccharide receptor, toll-like receptor-4. Immunology 113(2):153–162
11. Zhang D, Chen L, Li S et al (2008) Lipopolysaccharide (LPS) of Porphyromonas gingivalis induces IL-1beta, TNF-alpha and IL-6 production by THP-1 cells in a way different from that of Escherichia coli LPS. Innate Immun 14(2):99–107
12. He G, Karin M (2011) NF-κB and STAT3 – key players in liver inflammation and cancer. Cell Res 21(1):159–168
13. Cao S, Liu M, Sehrawat TS et al (2021) Regulation and functional roles of chemokines in liver diseases. Nat Rev Gastroenterol Hepatol 18(9):630–647
14. De Vito P, Incerpi S, Pedersen JZ et al (2011) Thyroid hormones as modulators of immune activities at the cellular level. Thyroid 21(8): 879–890
15. Rozing MP, Westendorp RGJ, Maier AB et al (2012) Serum triiodothyronine levels and inflammatory cytokine production capacity. Age (Dordr) 34(1):195–201
16. Mooradian AD (2019) Age-related resistance to thyroid hormone action. Drugs Aging 36(11):1007–1014
17. Papanicolaou DA (2000) Euthyroid sick syndrome and the role of cytokines. Rev Endocr Metab Disord 1(1–2):43–48
18. van der Spek AH, Jim KK, Karaczyn A et al (2018) The thyroid hormone inactivating type 3 deiodinase is essential for optimal neutrophil function: observations from three species. Endocrinology 159(2):826–835
19. van Zeijl CJJ, Surovtseva OV, Kwakkel J et al (2014) Thyrostimulin deficiency does not alter peripheral responses to acute inflammation-induced nonthyroidal illness. Am J Physiol Endocrinol Metab 307(6):E527–E537
20. Kwakkel J, Chassande O, van Beeren HC et al (2010) Thyroid hormone receptor {alpha} modulates lipopolysaccharide-induced changes in peripheral thyroid hormone metabolism. Endocrinology 151(4):1959–1969
21. Boelen A, Wiersinga WM, Fliers E (2008) Fasting-induced changes in the hypothalamus-pituitary-thyroid axis. Thyroid 18(2):123–129
22. Beigneux AP, Moser AH, Shigenaga JK et al (2000) The acute phase response is associated with retinoid X receptor repression in rodent liver. J Biol Chem 275(21):16390–16399
23. Lu B, Moser AH, Shigenaga JK et al (2006) Type II nuclear hormone receptors, coactivator, and target gene repression in adipose tissue in the acute-phase response. J Lipid Res 47(10):2179–2190

Chapter 5

Analysis of Apoptosis by Thyroid Hormone Induction

Darío Cilleros-Rodríguez and Marina Lasa

Abstract

Apoptosis is a type of programmed cell death which can be induced by thyroid hormone in pituitary and other cell models. This process is characterized by several biochemical changes, including modifications in plasma membrane lipid composition, mitochondrial membrane dysfunction, and DNA fragmentation, among others. These cellular alterations can be measured using western blotting and flow cytometry techniques. Here we describe the method to detect cleavage of poly-ADP ribose polymerase by western blotting, as well as different flow cytometry-based methods to quantify the sub-G1 hypodiploid cell population, the redistribution of phosphatidylserine to the outer leaflet of plasma membrane, and the loss of the mitochondrial inner membrane potential.

Key words Thyroid hormone, Apoptosis, Western blotting, Flow cytometry

1 Introduction

Apoptosis is a type of programmed cell death that is based on eliminating damaged, dysfunctional, or unnecessary cells in order to ensure the correct functioning of cell populations and tissues. Therefore, deregulation of the apoptotic process, whether due to increase or defect, has been associated with the appearance of pathologies such as cancer, as well as autoimmune or neurodegenerative diseases [1]. The apoptotic cell undergoes very characteristic biochemical changes, which can be easily monitored [2]. For example, lipid externalization occurs, in which phosphatidylserine is exposed on the external side of the cell membrane, favoring the recognition and phagocytosis of the apoptotic cell [3]. Additionally, fragmentation of cellular DNA also appears due to the activity of specific endonucleases, which cleave genomic DNA between nucleosomes [4]. These processes are mediated by the activation of caspases, cysteine proteases which are synthesized as inactive pro-enzymes and that, after their proteolysis by other caspases, act specifically on target proteins, giving rise to the entry of cell into apoptosis [5].

Constanza Contreras-Jurado (ed.), *Thyroid Hormones: Methods and Protocols*, Methods in Molecular Biology, vol. 2876,
https://doi.org/10.1007/978-1-0716-4252-8_5,

In mammals, apoptosis is initiated by two interconnected cascades, the intrinsic and extrinsic pathways, which converge in the execution pathway [2]. The intrinsic pathway, also known as the mitochondrial pathway, is activated in response to DNA damage, alterations in cellular organelles, or other intracellular signals. This pathway is regulated by the Bcl-2 family, which induces the loss of the mitochondrial inner membrane potential ($\Delta\psi m$) and the release to the cytosol of different pro-apoptotic proteins, such as cytochrome c, leading to the activation of caspase 9 [1]. Alternatively, the extrinsic pathway is mediated by the activation of TNFR (tumor necrosis factor receptor), which interacts with the inactive form of caspase 8, promoting its activation [6]. The intrinsic and extrinsic pathways converge in the execution pathway, considered the final phase of the apoptotic process. This pathway is characterized by the activation of execution caspases (caspases 3, 6, 7), which have previously been activated by the initiator caspases (caspases 2, 8, 9). Among the execution caspases, the most important is caspase 3, which carries out important functions in apoptosis, such as reorganization of the cytoskeleton and the formation of apoptotic bodies [5]. Likewise, it also induces one of the main hallmarks of apoptosis, the cleavage of PARP (poly-ADP ribose polymerase), a protein involved in the recognition and attraction of damaged DNA repair systems [7].

Several methods are currently used to measure apoptosis in mammalian cells, some of them based on flow cytometry. For example, cell death can be determined by quantifying the sub-G1 hypodiploid cell population after propidium iodide (PI) staining, since this compound is a DNA dye that functions as a plasma membrane permeability indicator and can be used to follow DNA fragmentation. Hence, in this assay, the reduced DNA staining of apoptotic cells gives rise to a sub-G1 peak. In addition, PI can be combined with Annexin V, which is used to mark the redistribution of phosphatidylserine to the outer leaflet of the plasma membrane. In this case, different cell populations can be distinguished: viable (annexin V and PI double-negative cells), early apoptotic (annexin V single-positive cells), late apoptotic (annexin V and PI double-positive cells), and necrotic (PI single-positive cells). A third method to detect apoptosis by flow cytometry is based on the use of tetramethyl rhodamine methyl ester (TMRM), which is a cationic fluorescent dye that accumulates in mitochondria showing an intact $\Delta\psi m$. Therefore, this method allows us to discriminate cells with intact mitochondria from those damaged in the apoptotic process (which do not accumulate the TMRM reagent and show lower fluorescence values). This method is useful to determine whether apoptosis occurs through the intrinsic pathway. On the other hand, apoptosis can be also detected by a western blotting technique based on the analysis of PARP cleavage, providing information about the executing pathway.

All these methods have been employed to analyze the effect of thyroid hormone (T3) on apoptosis in different cell models. Thus, our group has demonstrated that this hormone induces apoptosis in different serum-starved pituitary cells through diverse mechanisms [8]. In fact, T3 increases the percentage of sub-G1 hypodiploid cells, as well as the percentages of cell populations that display early or late apoptosis [8]. Furthermore, T3 decreases TMRM fluorescence, indicating that it reduces mitochondrial inner membrane potential. Finally, T3 increases PARP cleavage, which is directly related to the activation of the caspase cascade characteristic of the apoptosis process [8]. These data are consistent with several reports in diverse cell types, showing that this hormone augments the early apoptotic cell population [9], induces a reduction of mitochondrial transmembrane potential [10], as well as increases caspase activities [11] and PARP cleavage [9, 12].

2 Materials

2.1 Cell Culture and Treatments

1. GH_4C_1, GH_3B_6, and GH_1 cell lines are models of different pituitary tumors. Cell culture media (DMEM, RPMI, or Ham's F10) are supplemented with either fetal bovine serum (FBS) or horse serum (HS). Culture GH_4C_1, GH_3B_6, or GH_1 cells in DMEM supplemented with 10% FBS, Ham's F10 supplemented with 10% HS, or RPMI supplemented with 2.5% FBS, respectively. Maintain all cell lines at 37 °C and 5% CO_2 in a humidified atmosphere and regularly test them to ensure the absence of mycoplasma.
2. AG1x8 resin and charcoal-stripped-FBS or -HS.
3. Washing buffer, Phosphate buffered saline, 1X PBS (137 mM NaCl, 2.7 mM KCl, 8.1 mM Na_2HPO_4, 1.8 mM KH_2PO_4, pH 7.4).
4. Solution to detach cells from culture plates, 1X TrypLE Express with EDTA, but without phenol red (*Gibco*).
5. Neutralization buffer, cell media supplemented with 10% serum, to inhibit enzymatic activity of trypsin.
6. Thyroid hormone (T3, *Sigma*), 1 mM stock solution in 0.04 N NaOH, stored in aliquots at −20 °C until use.
7. Staurosporine (STS, *Sigma*), 200 mM stock solution in dimethyl sulfoxide (DMSO), stored in aliquots at −20 °C until use.

2.2 Cell Extracts, Protein Quantification, and SDS/Polyacrylamide Gel (SDS/PAGE)

1. Lysis Buffer, 20 mM Tris (pH 7.4), 1 mM EDTA, 10% glycerol, 100 mM KCl, 1% Triton X-100, 0.3% β-mercaptoethanol, 5 mM $MgCl_2$. Store this buffer at 4 °C and the day of the experiment supplement an aliquot with 1 mM DTT, phosphatase inhibitors (5 mM NaF, 0.2 mM Na_3VO_4) and protease

inhibitors (1 mM PMSF, 1 μg/mL aprotinin, 1 μg/mL pepstatin, and 1 μg/mL leupeptin).

2. Bovine serum albumin (BSA), 1 mg/mL stock solution.
3. Bradford reagent (*BioRad*) in 1:4 dilution with pure and deionized water.
4. Laemmli Buffer 4X, 0.25 M Tris (pH 6.8), 8% sodium dodecyl sulfate (SDS), 0.05% bromophenol blue, 20% glycerol. Store this buffer at 4 °C and the day of the experiment supplement an aliquot with 5% β-mercaptoethanol.
5. Resolving gel Buffer, 1 M Tris buffer (pH 8.8), 40% acrylamide/bis-acrylamide solution (29:1), 10% SDS, 10% Ammonium persulfate (APS) solution in pure and deionized water, and N, N, N, N′-Tetramethyl-ethylenediamine (TEMED). Prepare this buffer immediately before polymerizing gel.
6. Stacking gel Buffer, same components as resolving gel buffer, except 1 M Tris buffer (pH 6.8). Prepare this buffer immediately before polymerizing gel.
7. SDS/PAGE Running Buffer, 10X Tris-Glycine (0.025 M Tris (pH 8.3) and 0.192 M glycine), 0.1% SDS.
8. Protein Markers for SDS/PAGE electrophoresis (*Nippon Genetics Europe*) as reference to determine molecular weight of protein samples. Store in aliquots at −20 °C to prevent repeated freeze-thaw cycles.

2.3 Immunoblotting and PARP Detection

1. Nitrocellulose blotting membrane 0.45 μM pore (*Protran, Amersham*).
2. Transfer Buffer, 0.025 M Tris, 0.192 M glycine, 0.1% SDS and 20% methanol. Store this buffer at 4 °C.
3. Tris-buffered saline (10X TBS) Solution, 1.5 M NaCl and 0.1 M Tris (pH 7.4).
4. TBS-T, TBS containing 0.1% Tween-20.
5. Blocking solution, 5% semi-skimmed milk in TBS-T.
6. Primary antibody, rabbit anti-PARP-1 (polyclonal antibody, sc-7150 *Santa Cruz Biotechnology*).
7. Secondary antibody, horseradish-peroxidase (HRP) conjugated to goat immunoglobulin (mainly, IgG) anti-Rabbit (polyclonal antibody, *DAKO*).
8. Immobilon Western Chemiluminescent HRP Substrate (*Millipore*), luminol solution and peroxide solution should be mixed in 1:1 ratio.
9. High-performance chemiluminescence ray films (*Amersham*), stored at room temperature (RT) in the dark to avoid direct light.

2.4 PI Staining, Fluorescein Isothiocyanate-Conjugated Annexin V (Annexin V-FITC)/PI Double Staining, TMRM Assay and Flow Cytometry Analysis

1. PI Fixing solution, 70% ice-cold ethanol, stored at 4 °C.
2. PI Staining solution, 1 mg/mL ribonuclease (RNase, *Sigma*) and 5 μg/mL PI (*Sigma*) dissolved in ice-cold 1X PBS. Solutions of RNase (stock 10 mg/mL) and PI (stock 1 mg/mL) are stored at −20 °C, and 4 °C protected from light, respectively.
3. Annexin V-FITC Apoptosis Detection Kit (*Merck Millipore*) [1]. Annexin V-FITC Staining solution, 10 μL/sample of Annexin V-FITC in 1X Binding Buffer supplemented with 0.5% FBS (*see* **Note 1**) [2]. PI Staining solution, 10 μL/sample of PI in 1X Binding Buffer supplemented with 0.5% FBS. All these reagents must be stored at 4 °C in the dark.
4. TMRM Staining solution, 50 nM TMRM reagent (*Invitrogen, ThermoFisher*) in ice-cold 1X PBS (*see* **Note 2**).
5. FACScalibur flow cytometer (*Beckton Dickinson*), equipped with a single 488 nm argon laser, and 530 nm, 585 nm, and > 650 nm filters for analysis in FL1, FL2, and FL3 channels, respectively. Excitation/Emission Peak of Annexin V-FITC is 498/517 nm, being measured with 525/40 nm filter, from FL1 channel. Excitation/Emission Peak of PI is 493/620 nm, being measured with 585/42 nm filter, from FL2 channel. Excitation/Emission Peak of TMRM is 548/574 nm, being measured with 585/42 nm filter, from FL2 channel.
6. FlowJo software, to analyze all data.

3 Methods

3.1 Cell Culture and Treatments

1. In all experiments, seed in 6-well plates an optimal number of cells to ensure control samples are confluent enough at the time of analyzing experiments. In general, the optimal range of 4–6 × 10^5 cells/well is needed during seeding to reach an appropriate confluency after three or four days.
2. Seed cells in media containing 0.1% AG1x8 resin and charcoal-stripped-FBS (for GH_4C_1 cells) or -HS (for GH_1 and GH_3B_6 cells). In both cases, resin and charcoal-stripping of sera is practiced in order to remove T3 traces.
3. Incubate cells during 24 h and treat them for 48 h with T3, either at 5 nM in GH_4C_1 and GH_1 cells, or at 100 nM in GH_3B_6 cells. In some experiments, STS is used as positive control to analyze apoptosis in pituitary cells [13] (*see* **Note 3**).

3.2 Detection of PARP Cleavage by Western Blotting

1. Place culture plates on ice, gently remove cell media and rinse cells twice with 1 mL chilled 1X PBS.

2. Once PBS is removed, add 100 μL/well of lysis buffer, and collect cell lysates in Eppendorf tubes using scrappers.
3. Keep lysates on ice for 10 min for a fully success lysis and clarify by centrifugation at 13000× *g* for 10 min at 4 °C.
4. Collect supernatants in clean Eppendorf tubes and freeze at −80 °C until use.
5. Perform quantification of protein concentration in samples by Bradford assay.
   5.1. Prepare a standard curve (using triplicates) with different concentrations of BSA (0–1.25-2.5-5 μg/μL) in a 96-well plate.
   5.2. Add 1 μL/well (using triplicates) of each problem sample. Ensure samples are thawed on ice and well mixed after briefly spinning them at 13000× *g*, 4 °C.
   5.3. Add 200 μL/well of diluted Bradford 1:4 and mix gently the plate up-to-down and left-to-right.
   5.4. Measure absorbance at 595 nm in a spectrophotometer (*VersaMax, Molecular Devices*). Calculate the concentration (μg/μL) of each sample using the BSA standard curve as a reference.
6. Prepare samples for loading into SDS/PAGE. Take the appropriate volume (μL) of each sample to get an amount of total protein of 40–60 μg for cleaved PARP visualization.
7. Add the corresponding volume of 4X Laemmli Buffer to reach a final 1X in each sample and mix well. All tubes should be RT, out of ice, preventing precipitation of SDS from the Laemmli buffer.
8. Spin samples at 13000× *g*, boil them at 95 °C for 5 min and centrifuge at 13000× *g*. Samples are ready to directly load and run in SDS/PAGE electrophoresis or can be frozen at −20 °C.
   8.1. Prepare an 8% SDS/PAGE (*see* **Note 4**).
      8.1.1. Resolving Gel: Mix 2 mL 40% acrylamide solution, 3.78 mL 1 M Tris buffer (pH 8.8), 100 μL 10% SDS and 4 mL ultrapure and deionized water to reach a final volume of 10 mL of solution. Shake reagents by vortexing and add 100 μL 10% APS and 4 μL TEMED to initiate reaction of gel polymerization. Drop around 7–8 mL of the content to the glasses where SDS/PAGE is prepared. Allow space for stacking gel and gently overlay up to top of the glass with isobutanol. Incubate RT until gel is formed. Once gel has polymerized, remove isobutanol as much as possible to avoid leaving traces in the solution.

8.1.2. Stacking Gel: Mix 1.3 mL 40% acrylamide solution, 1.25 mL 1 M Tris-HCl buffer (pH 6.8), 100 μL 10% SDS and 7.2 mL ultrapure and deionized water to reach a final volume of 10 mL of solution. Vortex solution and add 100 μL 10% APS and 10 μL TEMED. Drop solution up to the top of the glass and insert a 10-well gel comb immediately without introducing air bubbles. Incubate RT until stacking gel is formed (*see* **Note 5**).

8.2. Prepare 1 L of running buffer, put gel glasses in their package, pour running buffer inside the whole surface of the package ensuring gels are fully covered, and gently remove combs.

8.3. Load firstly protein markers in their appropriate lanes and then load the desired volume of each sample prepared as previously described (*see* **Note 6**).

8.4. Run electrophoresis at 100 V during 120–150 min. Let gel run until dye front reaches the bottom of the gel.

9. Perform immunoblotting:

9.1. Following electrophoresis, open gel-containing glasses with the use of a spatula and discard the stacking gel. The resolving gel will remain on one of the glass plates. Rinse the gel with water and place carefully to a container with western blot transfer buffer.

9.2. Cut a nitrocellulose membrane to the size of the gel and immerse in transfer buffer during 5–10 min before starting wet electroblotting (*see* **Note 7**).

9.3. Prepare gel-membrane-Whatman sandwich in a specific cassette (*BioRad*), lifting the gel from the glass plate and place it on nitrocellulose membrane. Covering the sandwich, put a total of 4 Whatman filter papers cut to the shape of gel and membrane and protect them with sponges in each limit of sandwich. The order is as follows: first, put two Whatman filter papers on the sponge of the cassette; then add nitrocellulose membrane on it; next, place gel on membrane; after that, cover the gel adding two more Whatman filter papers; and finally, put the other sponge of the cassette. Try to avoid the presence of bubbles inside the whole sandwich.

9.4. Place cassette in the tray with the transfer apparatus and fully cover with transfer buffer. Set wet transfer at 400 mA during 1 h at 4 °C (*see* **Note 8**).

9.5. During transfer, prepare 1 L TBS-T solution, dissolving 100 mL 10X TBS in 900 mL ultrapure and deionized water, and then add 1 mL Tween-20 to reach a final concentration of 0.1% detergent. Be sure Tween-20 is well mixed in TBS shaking solution for several minutes.

9.6. Check that transfer has occurred successfully by incubating the membrane for 10 min in a 0.5% Ponceau red solution in 1% acetic acid, and rinse the membrane in TBS-T for 10 min.

9.7. Prepare blocking solution: weigh a desired amount of powder milk and mix in TBS-T to make a 5% milk solution. After 10 min washing the membrane, place it into the blocking solution. Incubate solution overnight at 4 °C with gentle rocking (*see* **Note 9**).

9.8. Prepare a 1:2000 dilution of rabbit anti-PARP-1 antibody in blocking solution and incubate membrane for 2 h at RT (*see* **Note 10**). After incubation, collect antibody in a tube and keep it at −20 °C for the next use.

9.9. Wash membrane 5x with TBS-T for 10 min each time.

9.10. Prepare a 1:5000 dilution of HRP conjugated to goat immunoglobulin anti-Rabbit in blocking solution and incubate membrane for 1 h at RT.

9.11. Rinse membrane 5x in TBS-T for 10 min each time.

9.12. Prepare 1 mL of immobilon western chemiluminescent HRP mix and incubate membrane for 5 min.

9.13. Gently remove rests of reagent, place membrane in cassette and use high resolution chemiluminescence X-ray films to detect bands corresponding to full-length (116 kDa) and cleaved PARP (89 kDa) (Fig. 1).

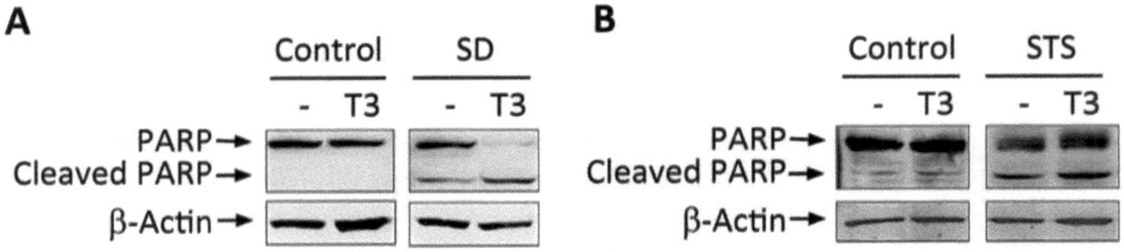

Fig. 1 Thyroid hormone induces apoptosis by promoting PARP cleavage. (**a**) GH_4C_1 cells were incubated for 48 h in medium supplemented with 10% (control) or 0.1% (SD) charcoal-stripped-FBS in the presence or absence of T3 (5 nM). Western blotting was performed to determine the expression of PARP, ensuring equal protein loading with β-actin. (**b**) Cells were incubated for 48 h in medium supplemented with 10% charcoal-stripped-FBS in the presence or absence of T3 (5 nM), and staurosporine (STS, 500 nM) was added for the last 24 h. Western blotting was performed to determine the expression of PARP, ensuring equal protein loading with β-actin. This figure has been modified from Chiloeches et al. [8]

3.3 Quantification of the Sub-G1 Hypodiploid Cell Population by PI Staining

1. Collect cell media in specific tubes for flow cytometry analysis, to study death rates also in cells detached from the plates.
2. Rinse with 1 mL/well 1X PBS to remove rests of cells or media with serum that could inhibit trypsin activity.
3. Add 300 μL 1X TrypLE Express and incubate for 5 min at 37 ° C, to ensure cells are successfully detached from culture plates.
4. Add 1 mL of fresh cell medium supplemented with serum and resuspend the content of each well several times to optimize cell detachment and dispersion from cell aggregations (*see* **Note 11**).
5. Collect well content in the tubes and immediately centrifuge samples at 390 x *g* for 5 min. Ensure to use a previously chilled centrifuge at 4 °C.
6. Discard supernatant, resuspend and rinse cell pellets in 500 μL ice-cold 1X PBS. Centrifuge cells again at same conditions mentioned above.
7. Add 500 μL of PI Fixing solution and incubate at 4 °C for at least 30 min, to ensure cell permeabilization (*see* **Note 12**).
8. Discard PI Fixing solution by centrifugation and wash cells twice in ice-cold 1X PBS.
9. During permeabilization, prepare PI Staining solution, mix well by gently vortexing and protect the solution from light.
10. After washing steps, centrifuge samples and incubate for 30 min in 500 μL PI Staining solution at 37 °C in the dark (*see* **Note 13**).
11. Analyze samples in FACScalibur flow cytometer (*Beckton Dickinson*).
    11.1. Select cell population, since changes in cellular morphology can be detected by measuring alterations in the way cells scatter the light of the laser beam in a flow cytometer. To assess this, set a gate for viable cells and early stages dead (apoptotic) cells on the FSC (forward scatter) vs. SSC (side scatter) dot plot. Viable cells will show an increase in FSC and SSC, while early apoptotic cells will show a slight decrease in FSC and an increase in SSC.

    11.2. Once the cell population is selected (without cell debris), establish a new gate to analyze cell singlets, discarding groups of cells or aggregates that cannot be adequately analyzed by flow cytometer. To create this gate, use the FSC-A vs. FSC-H dot plot, draw a region that includes the population diagonally, and select exclusively the population of cells within this gate for further analysis. Keep in mind that both axes of the dot plot represent the size of cell population (FSC), there being a positive linear correlation between the parameters corresponding to the area (FSC-A) and the height (FSC-H) of the cell population. Therefore, any event that shows higher values of area in population size (FSC-A) with similar values of height (FSC-H) must be discarded from the gate of this dot plot, since these events do not correspond to cell singlets, but rather to cell clumps (*see* **Note 14**).

11.3. Make a histogram, showing the counts and the ranges of fluorescence collected by using FL2 channel (585/42 nm filter). Then, analyze every cell population by setting the highest peak as G1 phase-cells, as reference in control samples. Sub-G1 hypodiploid cell population is represented as the first peak that appears in the histogram before the G1 peak, due to the near lack of PI staining of apoptotic and dead cells, which shows short and fragmented cellular DNA (**Fig. 2**).

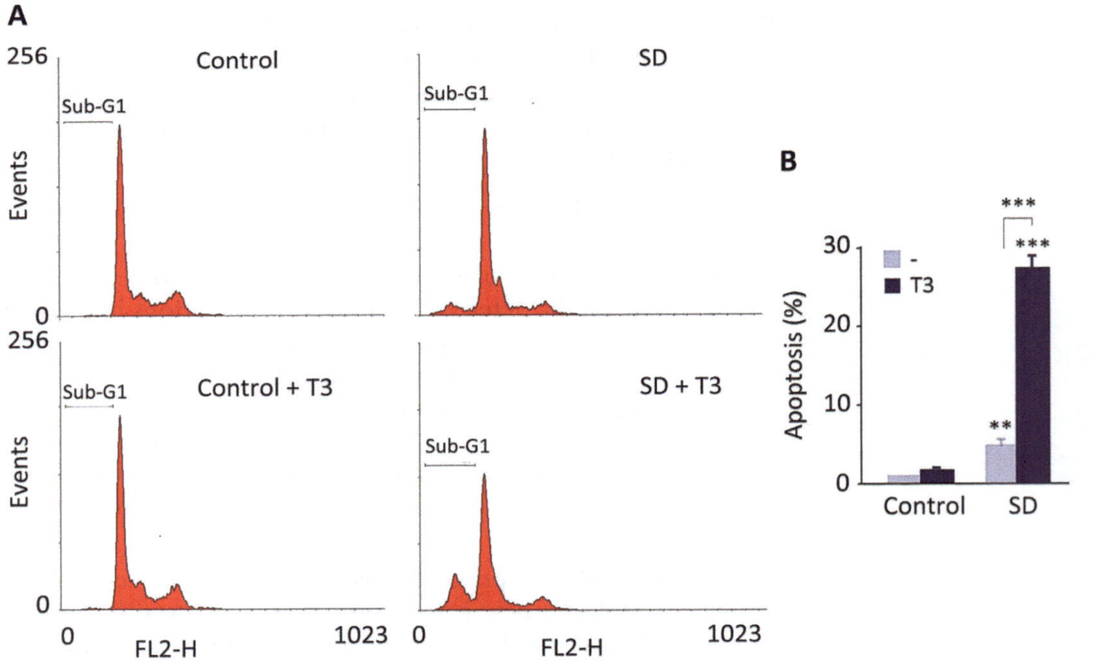

Fig. 2 Thyroid hormone induces apoptosis by increasing sub-G1 hypodiploid cell population. (**a**) GH_4C_1 cells were incubated for 48 h in medium supplemented with 10% (control) or 01.% (SD) charcoal-stripped-FBS in the presence or absence of T3 (5 nM). Apoptosis was determined in PI-stained cells and analyzed by flow cytometry. (**b**) Apoptosis values represent the mean ± SD of three independent experiments performed as in *(a)*. This figure has been modified from Chiloeches et al. [8]

3.4 Quantification of Apoptosis Based on the Redistribution of Phosphatidylserine to the Outer Leaflet of Plasma Membrane, by Annexin V-FITC/PI Double-Staining

Although this method consists of double staining to analyze several parameters of cell population, using two different channels from flow cytometer at the same time, some steps are similar to those belonging to the procedure used for the analysis of apoptosis after PI staining, described in Subheading 3.3. In fact, **steps 1** to **5**, as well as **steps 11.1** and **11.2** of both protocols are similar and, therefore, have been omitted in the description of this method.

1. Follow **steps 1–5** from Subheading 3.3.
2. Discard supernatant, resuspend cell pellets in 200 μL Annexin V-FITC staining solution, prepared as described in the Materials section (*see* **Note 15**), and incubate for 15 min at RT in the dark (*see* **Notes 13 and 16**).
3. Dilute Annexin V-FITC staining by adding 2 mL/sample of 1X Binding Buffer with 0.5% FBS, mix gently and centrifuge at 390 × g for 5 min at 4 °C.
4. Prepare PI staining solution and resuspend cell pellets in 100 μL 1X Binding Buffer without PI. When samples are ready to analyze in cytometer, add immediately 100 μL of PI staining solution to each sample and mix by gentle pipetting (*see* **Notes 13** and **16**).
5. Analyze samples in FACScalibur flow cytometer (*Beckton Dickinson*).
   - 5.1. Follow **steps 11.1** and **11.2** from Subheading 3.3.
   - 5.2. Using FL1 channel (525/40 nm filter) and FL2 channel (585/42 nm filter) and creating a FL1-A vs FL2-A dot plot, set a quadrant-shaped gate from singlet population, and make a histogram representing the counts and the measure of fluorescence of each sample collected by cytometer at this range of wavelengths in each channel. Different populations can be detected in cell samples, depending on whether they are positive for annexin V-FITC (measured with FL1 channel) and/or positive for PI (measured with FL2 channel). Thus, this method discriminates from viable cells (Annexin V-FITC$^-$/PI$^-$), early apoptotic cells (Annexin V-FITC$^+$/PI$^-$), late apoptotic cells (Annexin V-FITC$^+$/PI$^+$) and necrotic cells (Annexin V-FITC$^-$/PI$^+$) (Fig. 3).

3.5 Quantification of Loss of Mitochondrial Inner Membrane Potential by TMRM Staining

This protocol is very similar to the procedure used for the analysis of apoptosis after PI staining, described in Subheading 3.3. In fact, **steps 1** to **6**, as well as **steps 11.1** and **11.2** of both protocols are similar and, therefore, have been omitted in the description of this method.

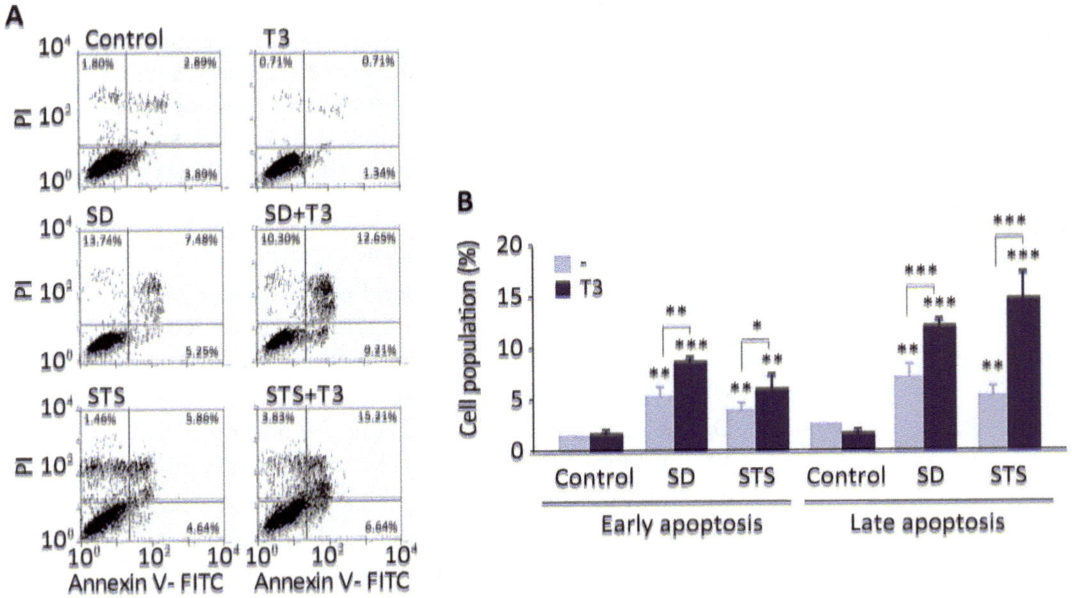

Fig. 3 Thyroid hormone induces early and late apoptosis. (**a**) GH_4C_1 cells were incubated for 48 h in medium supplemented with 10% (control) or 0.1% (SD) charcoal-stripped-FBS in the presence or absence of T3 (5 nM) or staurosporine (STS, 500 nM). Apoptosis was examined by the Annexin V/PI dual-staining assay, as detected by flow cytometry. (**b**) Early (Annexin V+/PI-) and late (Annexin V+/PI+) apoptosis values represent the mean ± SD of three independent experiments performed as in *(a)*. This figure has been modified from Chiloeches et al. [8]

1. Follow **steps 1–6** from Subheading 3.3.
2. Add 5 μL of the TMRM staining solution in the 500 μL of ice-cold 1X PBS previously used to resuspend cell pellets, in order to reach a final concentration of 50 nM TMRM. Mix by carefully shaking and incubate for 30 min at 37 °C in the dark (*see* **Note 13**).
3. Analyze samples in FACScalibur flow cytometer (*Beckton Dickinson*).
   3.1. Follow **steps 11.1** and **11.2** from Subheading 3.3.
   3.2. Select two new gates from the singlet population in the FL2-A (or FL2-H) vs. SSC-A (or SSC-H) dot plot. Make a histogram representing the counts and the measure of fluorescence collected by using FL2 channel (585/42 nm filter). With this procedure one can distinguish the cell population with low Δψm (apoptotic cells; left peak, cells with lower staining) from the one with higher Δψm (viable cells; right peak with higher staining) (*see* **Note 17**) (Fig. 4).

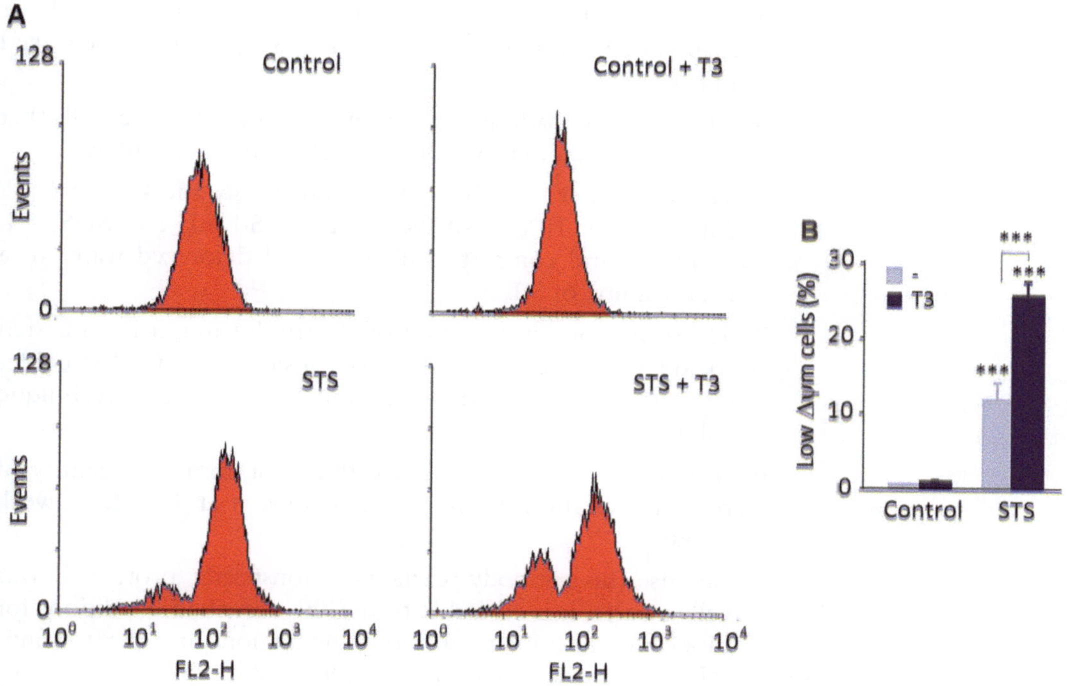

Fig. 4 Thyroid hormone induces the loss of the mitochondrial inner membrane potential. (**a**) GH_4C_1 cells were incubated for 48 h in medium supplemented with 10% (control) charcoal-stripped-FBS in the presence or absence of T3 (5 nM) or staurosporine (STS, 500 nM). Apoptosis was examined by TMRM staining assay, as detected by flow cytometry. (**b**) Values corresponding to the loss of the mitochondrial inner membrane potential represent the mean ± SD of three independent experiments performed as in *(a)*

4 Notes

1. The optimal rate of concentration for Annexin V-FITC staining is 0.25–2.5 μg/mL. Binding Buffer from Kit is 20X, so dilute buffer up to 1X adding ultrapure and deionized water; and next, supplement with 0.5% FBS.
2. The TMRM stock solution must be previously prepared at 5 mM in DMSO. Store in aliquots at −20 °C protected from light.
3. In some experiments, cells were treated with 500 nM STS. In these cases, STS was added for the last 24 h, while control cells were incubated with the same volume of the vehicle used to dissolve this compound.
4. To properly PARP visualization, it is imperative to use an 8% SDS/PAGE. Two bands corresponding to full-length PARP (116 kDa) and cleaved PARP (89 kDa), as a substrate of caspase 3, will be detected. Due to the high amounts of full-length PARP compared to those of cleaved PARP in total lysates, and the small difference in their molecular weights, it is essential to prepare an 8% SDS/PAGE to discriminate both bands.

5. Once SDS/PAGE is formed, it can be used immediately or maintained in fridge at 4 °C. In this case, gel should be sinked in running buffer.
6. Check before loading there are no rests of gel inside wells that could avoid adding the volume of the samples properly.
7. Transfer buffer must be prepared previously and kept at 4 °C. Mix 500 mL 10X Tris-Glycine buffer, 50 mL 10% SDS, 1 L methanol and complete with pure and deionized water to a total volume of 5 L.
8. To ensure correct operation of electroblotting, it is suggested to add a package with an ice block inside the tank. The colder electroblotting is carried out, the better transfer technique will work.
9. It is recommended to prepare the most accurate quantity of blocking solution during usage or store it at 4 °C for a week or less.
10. Because this antibody reacts with nonspecific proteins in our cell extracts, it is advisable to use the described conditions for blocking and primary antibody incubation, in order to minimize the appearance of nonspecific bands.
11. The minimal ratio to neutralize trypsin enzymatic activity is 1:3 (1X TrypLE Express:medium supplemented with serum).
12. Cells can also be permeabilized by incubating them for 16 h at −20 °C in a solution containing ethanol 70–80%. The colder the fixative solution and the longer the incubation time, the better the permeabilization and incorporation of the PI dye into the DNA.
13. To analyze cell autofluorescence, include cell pellets incubated in the absence of dye (PI, TMRM, or double Annexin V-FITC/PI) as negative controls. In these cases, resuspend cell pellets in the same volume of the vehicle used to dissolve the corresponding dye (ice-cold 1X PBS or 1X Binding Buffer supplemented with 0.5% FBS).
14. For PI staining, it is advisable to use directly a FL2-A vs. FL2-H dot plot to analyze all singlets from cell population that have incorporated PI dye.
15. In this case, instead of using ice-cold 1X PBS to dissolve stock staining reagents, it is necessary to employ a specific buffer, called Binding Buffer, which contains saline solutions such as NaCl, KCl, $MgCl_2$, and $CaCl_2$. The Ca^{2+} ions are essential to promote interaction between annexin V (conjugated to FITC fluorophore) and phosphatidylserines, located in outer leaflet of plasma membrane in apoptotic cells.

16. For double staining protocols, compensations are needed in order to prevent overlapping between both fluorescence signals in FL1 and FL2 channels. To assess this, prepare some samples known to be positive for Annexin V-FITC and even positive for PI (e.g., apoptotic cells), and dye them with only one of the staining solutions, which is called "Single color Annexin V-FITC" and "Single color PI". This also enables to set negative cell population near "0" in dot plot axis; and positive cell population, inside the scale of fluorescence measured in dot plot axis.
17. For more details regarding the setting of gates in TMRM staining method, see more results in [14, 15].

References

1. Singh R, Letai A, Sarosiek K (2019) Regulation of apoptosis in health and disease: the balancing act of BCL-2 family proteins. Nat Rev Mol Cell Biol 20(3):175–193
2. Elmore S (2007) Apoptosis: a review of programmed cell death. Toxicol Pathol 35(4): 495–516
3. Crimi M, Esposti MD (2011) Apoptosis-induced changes in mitochondrial lipids. Biochim Biophys Acta 1813(4):551–557
4. Matassov D, Kagan T, Leblanc J, Sikorska M, Zakeri Z (2004) Measurement of apoptosis by DNA fragmentation. Methods Mol Biol 282: 1–17
5. Ghavami S, Hashemi M, Ande SR, Yeganeh B, Xiao W, Eshraghi M, Bus CJ, Kadkhoda K, Wiechec E, Halayko AJ, Los M (2009) Apoptosis and cancer: mutations within caspase genes. J Med Genet 46(8):497–510
6. Cavalcante GC, Schaan AP, Cabral GF, Santana-da-Silva MN, Pinto P, Vidal AF et al (2019) A Cell's fate: an overview of the molecular biology and genetics of apoptosis. Int J Mol Sci 20(17):4133–4153
7. Soldani C, Scovassi AI (2002) Poly (ADP-ribose) polymerase-1 cleavage during apoptosis: an update. Apoptosis 7(4):321–328
8. Chiloeches A, Sanchez-Pacheco A, Gil-Araujo B, Aranda A, Lasa M (2008) Thyroid hormone-mediated activation of the ERK/dual specificity phosphatase 1 pathway augments the apoptosis of GH4C1 cells by down-regulating nuclear factor-kappaB activity. Mol Endocrinol 22(11):2466–2480
9. Sar P, Peter R, Rath B, Das Mohapatra A, Mishra SK (2011) 3, 3′5 Triiodo L thyronine induces apoptosis in human breast cancer MCF-7 cells, repressing SMP30 expression through negative thyroid response elements. PLoS One 6(6):e20861
10. Mihara S, Suzuki N, Wakisaka S, Suzuki S, Sekita N, Yamamoto S, Saito N, Hoshino T, Sakane T (1999) Effects of thyroid hormones on apoptotic cell death of human lymphocytes. J Clin Endocrinol Metab 84(4):1378–1385
11. Yamada-Okabe T, Satoh Y, Yamada-Okabe H (2003) Thyroid hormone induces the expression of 4-1BB and activation of caspases in a thyroid hormone receptor-dependent manner. Eur J Biochem 270(14):3064–3073
12. Bolf EL, Gillis NE, Davidson CD, Rodriguez PD, Cozzens L, Tomczak JA, Frietze S, Carr FE (2020) Thyroid hormone receptor Beta induces a tumor-suppressive program in anaplastic thyroid cancer. Mol Cancer Res 18(10): 1443–1452
13. Fombonne J, Reix S, Rasolonjanahary R, Danty E, Thirion S, Laforge-Anglade G, Bosler O, Mehlen P, Enjalbert A, Krantic S (2004) Epidermal growth factor triggers an original, caspase-independent pituitary cell death with heterogeneous phenotype. Mol Biol Cell 15(11):4938–4948
14. Gil-Araujo B, Toledo Lobo MV, Gutierrez-Salmeron M, Gutierrez-Pitalua J, Ropero S, Angulo JC, Chiloeches A, Lasa M (2014) Dual specificity phosphatase 1 expression inversely correlates with NF-kappaB activity and expression in prostate cancer and promotes apoptosis through a p38 MAPK dependent mechanism. Mol Oncol 8(1):27–38
15. Martinez-Martinez D, Soto A, Gil-Araujo B, Gallego B, Chiloeches A, Lasa M (2019) Resveratrol promotes apoptosis through the induction of dual specificity phosphatase 1 and sensitizes prostate cancer cells to cisplatin. Food Chem Toxicol 124:273–279

Chapter 6

Immunophenotyping of Leukocytes in Brain in Hypothyroid Mice

Ángela Sánchez

Abstract

Hypothyroidism, characterized by inadequate production of thyroid hormones, and malaria, a mosquito-borne infectious disease caused by *Plasmodium* parasites, are significant health concerns worldwide. Understanding the interplay between these two conditions could offer insights into their complex relationship and potential therapeutic strategies. To induce hypothyroidism, pharmacological inhibition of thyroid hormone synthesis was employed. Subsequently, mice were infected with *Plasmodium berghei ANKA* to simulate cerebral malaria infection. It needs to monitor the progression of the disease in male mice before it can identify infiltrating immune system populations of interest in the brain by multiparametric techniques such as flow cytometry.

Key words Hypothyroidism, Cerebral malaria, *Plasmodium berghei ANKA*, Flow cytometry, Leukocytes

1 Introduction

Globally in 2022, there were almost 250 million estimated malaria cases in 85 malaria endemic countries, an increase of five million cases compared with 2021 [1]. Malaria or paludism is a devastating infection that constitutes a serious health problem. It is one of the most relevant worldwide parasitic diseases due to its high mortality, morbidity, and socioeconomic impact that it causes in the human population.

A unicellular protozoan from the genus *Plasmodium* is the etiological agent in mammals. In humans, the infection is transmitted through dipteran vectors such as *Anopheles, Aedes*, or *Culex* [2]. Parasites enter the bloodstream via mosquito bite and spend several parts of their life cycle inside humans which infect and reproduce inside the liver and red blood cells [3].

Malaria with brain damage or cerebral malaria (CM) is the most lethal form of disease [4, 5] and it has a devastating effect on

Constanza Contreras-Jurado (ed.), *Thyroid Hormones: Methods and Protocols*, Methods in Molecular Biology, vol. 2876, https://doi.org/10.1007/978-1-0716-4252-8_6,

children under the age of 5 [6]. Globally, although the cases of infections have decreased in the last decade, the mortality rate of cerebral malaria is close to 10% in children under 5 years old [4, 7]. His devastation effects have been reported both at the organ and molecular levels. CM is characterized by blood-brain barrier disruption followed by an increase of intracranial pressure with compression of cerebral vessels due to brain swelling [8]. This scenario causes altered consciousness, seizures, total or partial paralysis, and coma, which finally conduce to death [2, 4, 7, 9]. In humans, *Plasmodium falciparum* directly hemolyzes red blood cells and in children, the parasites have been reported to have deleterious effect on the gastrointestinal system and the brain between others [10–12].

The most widely experimental model used to study malaria progress is the infection with *Plasmodium berghei ANKA* (PbA) in C57BL/6 mice [13], although this strain is not able to develop cerebral malaria in all individuals. It has been demonstrated that the disease development depends greatly on the parasite-host relationship [14, 15]. In any case, infected erythrocytes are inoculated to healthy mice that can recapitulate the pathological symptoms of cerebral malaria such as ataxia, weakness, hunching, seizures, paralysis, coma, and death within 5 to 7 days [16] and the process has numerous similarities with the pathological process caused by *P. falciparum* in humans [15, 17]. Also at the cellular level, the higher cytokine levels and increase of ICAM-1, VCAM-1, and P-selectin with the activation of the endothelia brain are very similar to the effects in humans [14, 18].

Recently reports indicated that specific parasite phenotypes are focused on affecting the small blood vessels in the brain, leading to problems with the endothelial cells and resulting in brain swelling [4]. The blood-brain barrier breakdown is crucial for this pathology development. This fact has been observed in the infection caused by PbA and *P. falciparum* in mice and humans respectively by the accumulation of infected erythrocytes and leukocytes in brain tissue (CD8+ T lymphocytes, macrophages, and monocytes) [17–19].

Reports have shown that macrophages and monocytes produce cytokines such as TNF-α, IL-1a, or IL-1b, which activate brain endothelial cells during the infectious process. Additionally, endothelial cells release other cytokines such as CCL2, CCL4, CXCL4, and CXCL8, which facilitate the recruitment of T lymphocytes, including CD4+ and CD8+ cells, to the brain [16, 20, 21]. CD8+ T lymphocytes indirectly promote the endothelial cell apoptosis facilitating the blood-brain barrier breakdown [22, 23]. A vital population of Ly6C+ monocytes has been identified as crucial for recruiting CD8+ T lymphocytes to the central nervous system and for cerebral malaria development [24]. During malaria, the immune responses are triggered when resident splenic dendritic cells (DCs) are exposed to parasite nucleic acids. This exposure prompts the production of IFNγ by CD4+ and CD8+ T lymphocytes and induces the differentiation of inflammatory monocytes into monocyte-derived dendritic cells (Mo-DCs). These Mo-DCs play a significant role in initiating the neuroinflammation typically associated with cerebral malaria by expressing the cytokine CCR5 [25].

The immune system can identify and eradicate the invading pathogens and the host organism can also defend itself from infectious diseases by reducing the adverse effects of infections on host fitness. This capacity to tolerate the presence of a pathogen represents a unique host defense approach, is specific and depending on the nature of the pathogen, and is not exclusive of the rest of the immune system mechanisms [26, 27]. In this context, recent studies demonstrate that hypothyroidism confers tolerance to cerebral malaria. Modulating the host's metabolism to protect tissue from damage induces tolerance to infections increasing survival and reducing the function of thyroid gland activity during malaria infection could aggravate the outcome of the disease, but it could also constitute a protective mechanism [8]. Thyroid hormones play a crucial role in the regulation of certain key metabolic enzymes, both in peripheral tissues and in the brain. It has been described that T3 controls energy consumption via regulation of mTORC1 activity in the arcuate nucleus of the hypothalamus [28]. Growing evidence has accumulated suggesting the ability of thyroid hormones to modulate hematopoiesis and the immune response in both humans and animals. It has been reported the involvement of thyroid receptor α in the maturation of erythrocyte progenitors. Disruption of thyroid balance significantly influences the development of the red blood cell series and anemia is a common symptom among patients suffering from hypothyroidism [29–34].

The purpose of this chapter is to detail first how mice become hypothyroidism and then develop the own pathology of cerebral malaria infection. In second hand, to explain how infiltrated hematopoietic cells in the brain were identified by flow cytometry analysis since this technique has the potential to measure multiple parameters simultaneously and rapidly within mixed populations' cells.

2 Materials

2.1 Generation of Hypothyroid Mice

1. C57BL/6 male mice (*see* **Note 1**).
2. Propylthiouracyl.
3. Potassium perchlorate.

4. Control diet.
5. Potassium iodide.

2.2 Generation of Cerebral Malaria Infections

1. *Plasmodium berghei ANKA*.
2. Para-amino benzoic acid.
3. EDTA microtubes.
4. Sterile or ethanol-cleaned surgical instruments.
5. Methanol.
6. Wright's solution.
7. Glass slides.
8. Microscope with x100 objective.
9. MiliQ water.

2.3 Isolation and Preparation of Mouse Brain Cells

1. Sterile or ethanol-cleaned dissection material.
2. 70% ethanol.
3. Mouse adult brain dissociation kit.
4. 70-μm cell strainer.
5. Brain gradient solution: 36% Percoll in DMEM-HEPES medium enriched with 0.4% Hanks′ balanced salt solution and 10% Fetal Bovine Serum (FBS).
6. ACK lysing buffer: 0.15 M ammonium chloride, 0.01 M potassium bicarbonate, and 0.1 mM disodium EDTA in distilled water. Adjust the pH to 7.2–7.4.
7. FACS buffer: PBS with 2% FBS and 5 Mm EDTA. Filtered.

2.4 Cell Surface Staining

All antibodies described herein are used at predetermined optimal concentrations (*see* **Note 2**).

1. FACS tubes: 12 × 75 mm polystyrene round bottom tubes.
2. Fluorochrome-labeled anti-mouse antibodies for flow cytometry against indicated cell surface markers: CD45, CD3, CD4, CD8 CD206a, CD11b, MHCII, F4/80. Clone and fluorochromes are described in Table 1 (*see* **Note 3**).
3. Microbeads Perfect-Count Microspheres.

2.5 Equipment and Supplies

1. Centrifuge.
2. Cytometer FACSCanto II (Becton and Dickinson).
3. FlowJo software.

Table 1
Fluorochrome-labeled anti-mouse antibodies for flow cytometry

Antibodies specificity	Fluorochrome	Clone
CD11b	APC-Cy7	MI-70
CD206a	BV 650	CO682C
CD3	APC	145-2C11
CD4	PE	GK1.5
CD45	PE-Cy7	30-F11
CD8	BV 450	53–6.7
F4/80	BV 421	T45–2342
MHCII	Super Bright 600	M5/114.15.2

3 Methods

3.1 Generation of Hypothyroid Mice

1. Wean male mice after 4–5 weeks from birth and feed them ad libitum with an iodine-deficient diet (<0,015 mg/kg) containing 0.15% of antithyroid drug propylthiouracil for four weeks.
2. At this moment, add to the drinking water potassium perchlorate (10 g/ liter). Change the drinking water every day.
3. To the group of control euthyroid mice, feed the animals with the control diet, the same diet without propylthiouracil but plus potassium iodide (1.15 mg/kg). Do not change both diets throughout the experiments.

3.2 Generation of Cerebral Malaria Infections

1. After 4 weeks of hypothyroid treatment, inoculate the mice with *P. berghei ANKA* (PbA) by intraperitoneal injection of $1x10^6$ infected red blood cells from infected donor mice as detailed below.
2. Euthanize donor mice according to a locally approved procedure (*see* **Note 1**).
3. Collect the blood into an EDTA tube through cardiac puncture and keep it at room temperature (RT).
4. Use one blood drop to make a smear in a glass slide and let dry at RT.
5. Fix the sample with methanol and let dry at RT.
6. Stain the sample by adding Wright's solution and incubate for 3 min at RT.
7. Clean the slide with MiliQ water until the staining solution is removed completely and let it dry at RT.

8. Observe the results under x100 objective and quantify the proportion of infected erythrocytes (*see* **Note 4**).
9. Calculate the blood volume necessary to inoculate $1x10^6$ infected red blood cells in 100 μL per mouse. Dilute the blood using sterile PBS 1X if it is necessary.
10. After inoculation, add 0.05% of para-aminobenzoic acid to the drinking water, which is needed to ensure the parasite life cycle. We define this time as day 0 post infection.
11. To monitor the parasite levels, extract a blood drop from the tail to make a smear every day starting the day 2–3 post infection. Follow the **steps 4** to **8** of this section to calculate the parasite levels.
12. Observe and monitor all mice groups daily to determine the clinical stage of cerebral malaria. Use the following criteria to classify the pathological symptoms of the disease [35].
    (0) Not symptoms.
    (1) Presence of piloerection.
    (2) Medium piloerection and mild ataxia.
    (3) Ataxia and paralysis.
    (4) Convulsions and coma.
13. Sacrifice the mice (*see* **Note 1**) when they show severe symptoms of the disease (stage 4).

3.3 Isolation and Preparation of Mouse Brain Cells

This subheading describes the dissection of the mouse brain and subsequent recovery of brain-infiltrated leukocytes.

1. Euthanize mice according to a locally approved procedure (*see* **Note 1**).
2. Disinfect the skin of the mouse with 70% ethanol using a spray bottle. Prepare 2 mL buffer per mouse to collect the brains.
3. Extract brain samples carefully just after animal sacrifice and mince them to process in the gentleMACS dissociator using the Adult Brain Dissociation Kit.
4. Filter samples through a 70-μm cell strainer to remove tissue fragments.
5. Centrifuge the cells in 10 ml of Brain gradient solution for 30 min at 800 g at RT. Use a centrifuge with no brake (*see* **Note 5**).
6. Collect the cell pellet at the bottom and wash it with 2 mL of FACS buffer.
7. Lyse the erythrocytes in 400 μL of ACK lysing buffer for 2 min (*see* **Note 6**).
8. Add 2 mL of FACS buffer.

9. Centrifuge the cells for 5 min at 300 g at RT for wash it.
10. Resuspend in an appropriate volume of FACS buffer.

3.4 Cell Surface Staining of Brain-Infiltrated Leukocytes

Each antibody is titrated before its use in order to adjust the optimal concentration. Prior to the actual staining of cells, antibody cocktails are prepared. The staining volumes indicated herein can be adjusted depending on cell numbers.

1. Separate an aliquot a small fraction of enriched cell to separate polypropylene tubes for FMO controls and sample.
2. Centrifuge 300 g for 5 min at RT.
3. Resuspend in 50 μL of FACS buffer.
4. Add an equal volume of antibody cocktail at double the optimal concentrations to reach a final optimal concentration of antibodies.
5. Incubate for 20 min in dark at RT.
6. Add 500 μL of FACS buffer to wash the cells.
7. Centrifuge 300 g for 5 min at RT.
8. Resuspend in appropriate volume to flow cytometry analysis.
9. One minute prior to cytometry analysis, add microbeads Perfect-Count Microspheres to measure the cells' absolute number.

3.5 Samples Acquisition

It is advisable to filter samples prior to cytometry analysis to minimize the risk of clogs. The flow cytometry data were acquired using FACSCanto II and analyzed with FlowJo software.

3.5.1 Compensation Procedures

1. Use an aliquot of unstained cells as negative control to set appropriate gains for forward and side scatter and the optimal flow rate.
2. For each fluorochrome antibody, use a single-color positive control to adjust the flow cytometer voltage setting and mix compensation as described below.
3. Prepare single-stained compensation controls by aliquoting 100 ul of cells ($0.5x10^6$ cells).
4. Add a volume of antibody at the optimal concentrations to reach a final optimal concentration of antibodies. Prepare as many single colors control as antibodies there are in the cocktail.
5. Follow the **steps 5** to **8** of Subheading 3.4.
6. Acquire all single-stained controls, including the negative control, and calculate compensation values across all included detectors according to software instructions.

7. For samples, acquire sufficient amounts of events to obtain statistically sound data with sufficient power.

3.5.2 Gate Setting Strategies

This subheading provides a strategy on how to set gates that define different immune system populations of interest. This discussion is based on the plots depicted in Fig. 1, which is described in a previous work [8]. Please note that most parameters are presented using a biexponential axis that uses alternative scaling of the lower end of the axis. This allows for presentation also of negative values and avoids events from "sticking to the axes," thereby maximizing the visualization of data. The three first steps are done for the exclusion of most cellular doublets:

1. Displaying SSC-area against FSC-area excludes smaller or larger particles such as unwanted cells and debris.
2. Displaying FCS-weight against FCS-area allows for the exclusion cellular doublets.
3. Displaying SSC-weight against SSC-area allows for the exclusion cellular doublets.
4. Displaying FCS-area against CD45 for select CD45-positive cells to continue.
5. Here, divide the strategy into 2 parts. First, to define T lymphocytes, and second to define monocyte-derived dendritic cells (Mo-DCs).

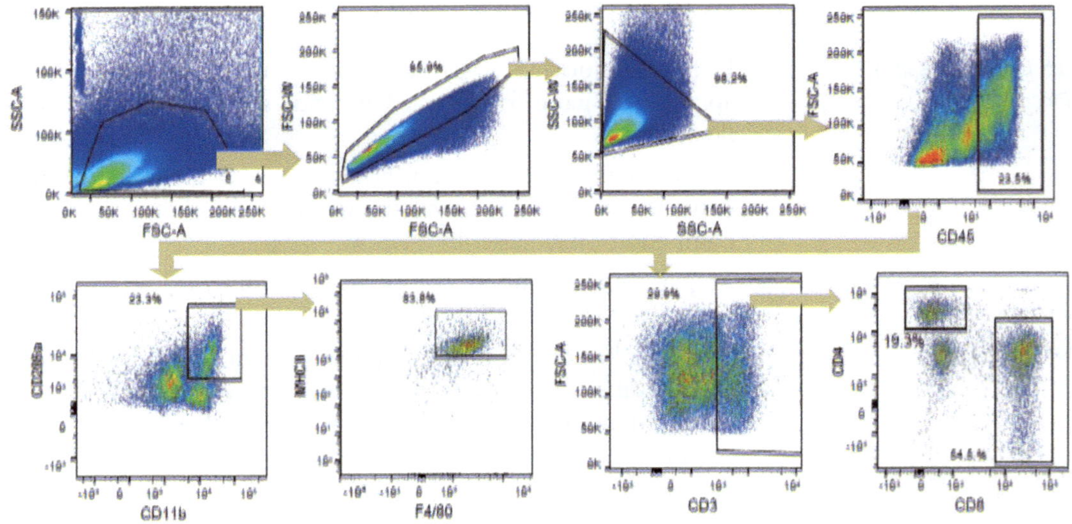

Fig. 1 Gating strategy of brain cell populations. On the right, single brain CD45+ cells were plotted for CD3, and CD45 + CD3+ cells were then split into subgroups based on their expression of CD4 or CD8. On the left, brain mono-DCs were identified by dividing single brain CD45+ cells on the basis of their positive expression of F4/80 and CD11b, which were further plotted for CD206a and MHCII

5.1. Plotting FCS-area against CD3 for select CD45 + CD3+ cells. Continue plotting CD4 against CD8 to obtain the following populations: CD45 + CD3 + CD4+ and CD45 + CD3 + CD8 + .

5.2. Plotting CD206a against CD11b for select double positive cells population. Continue plotting MHCII against F4/80 for select double positive cells population again. The Mo-DCs are identified such CD45 + CD206a + CD11b + MHCII+F4/80 + .

4 Notes

1. All protocols involving experimental animal work must receive approval from local ethics committees and adhere to national laws.
2. Although the company details the optimal concentration to use each antibody per million of cells, it is recommended to adjust the concentration for each cell suspension and cytometer. The intensity of lasers can change between different cytometers and also the lot of antibodies. For all of that, make a titration of each single antibody is desirable prior to carrying out the experiment.
3. Antibodies used herein are commercially available at one or several suppliers; we obtain most antibodies from BD Biosciences (San Jose, CA), eBioscience (San Diego, CA), or Bio-Legend (San Diego, CA).
4. You can use any software that allows image quantification. In this case, the samples were quantified using Fiji program.
5. To centrifuge with no brake is essential for no disruption of the different phases forming during the centrifugation step.
6. This step is not necessary if erythrocytes are not present.

References

1. WHO (2023) World malaria report 2023. World Health Organizartion. https://www.who.int/publications/i/item/9789240086173
2. Phillips M, Burrows J, Manyando C et al (2017) Malaria Nat Rev Dis Primers 3:17050
3. Ogbera AO, Anaba E (2021) Protozoa and endocrine dysfunction. In: Feingold KR, Anawalt B, Blackman MR et al (eds) Endotext. South Dartmouth (MA)
4. Wassmer SC, Grau GE (2017) Severe malaria: what's new on the pathogenesis front? Int J Parasitol 47(2–3):145–152
5. Good MF, Xu H, Wykes M et al (2005) Development and regulation of cell-mediated immune responses to the blood stages of malaria: implications for vaccine research. Annu Rev Immunol 23:69–99
6. Kwansa-Bentum H, Aninagyei E, Adedia D et al (2024) Elevation of free triiodothyronine

(fT3) levels by plasmodium falciparum independent of thyroid stimulating hormone (TSH) in children with uncomplicated malaria. J Clin Lab Anal 38(3):e25013
7. Cowman AF, Healer J, Marapana D et al (2016) Malaria: biology and disease. Cell 167(3):610–624
8. Rodriguez-Muñoz D, Sánchez Á, Pérez-Benavente S et al (2022) Hypothyroidism confers tolerance to cerebral malaria. Sci Adv 8(14): eabj7110
9. Nkumama IN, O'Meara WP, Osier FHA (2017) Changes in malaria epidemiology in Africa and new challenges for elimination. Trends Parasitol 33(2):128–140
10. Aninagyei E, Doku ET, Adu P et al (2018) Storage related haematological and biochemical changes in plasmodium falciparum infected and sickle cell trait donor blood. BMC Hematol 18(1):30
11. Agrawal D, Teach SJ (2006) Evaluation and management of a child with suspected malaria. Pediatr Emerg Care 22(2):127–133
12. Crawley J, Chu C, Mtove G et al (2010) Malaria in children. Lancet 375(9724): 1468–1481
13. Craig AG, Grau GE, Janse C et al (2012) The role of animal models for research on severe malaria. PLoS Pathog 8(2):e1002401
14. Lou J, Lucas R, Grau GE (2001) Pathogenesis of cerebral malaria: recent experimental data and possible applications for humans. Clin Microbiol Rev 14(4):810–820
15. Idro R, Jenkins NE, Newton CR (2005) Pathogenesis, clinical features, and neurological outcome of cerebral malaria. Lancet Neurol 4(12):827–840
16. Nishanth G, Schluter D (2019) Blood-brain barrier in cerebral malaria: pathogenesis and therapeutic intervention. Trends Parasitol 35(7):516–528
17. de Souza JB, Hafalla JC, Riley EM et al (2010) Cerebral malaria: why experimental murine models are required to understand the pathogenesis of disease. Parasitology 137(5): 755–772
18. Carvalho LJ (2010) Murine cerebral malaria: how far from human cerebral malaria? Trends Parasitol 26(6):271–272
19. Hunt NH, Grau GE, Engwerda C et al (2010) Murine cerebral malaria: the whole story. Trends Parasitol 26(6):272–274
20. Dunst J, Kamena F, Matuschewski K (2017) Cytokines and chemokines in cerebral malaria pathogenesis. Front Cell Infect Microbiol 7: 324
21. Chua CL, Brown G, Hamilton JA et al (2013) Monocytes and macrophages in malaria: protection or pathology? Trends Parasitol 29(1): 26–34
22. Renia L, Howland SW, Claser C et al (2012) Cerebral malaria: mysteries at the blood-brain barrier. Virulence 3(2):193–201
23. Schmid U, Stenzel W, Koschel J et al (2017) The deubiquitinating enzyme Cylindromatosis dampens CD8(+) T cell responses and is a critical factor for experimental cerebral malaria and blood-brain barrier damage. Front Immunol 8: 27
24. Pai S, Qin J, Cavanagh L et al (2014) Real-time imaging reveals the dynamics of leukocyte behaviour during experimental cerebral malaria pathogenesis. PLoS Pathog 10(7):e1004236
25. Hirako IC, Ataide MA, Faustino L et al (2016) Splenic differentiation and emergence of CCR5(+) CXCL9(+) CXCL10(+) monocyte-derived dendritic cells in the brain during cerebral malaria. Nat Commun 7:13277
26. Medzhitov R, Schneider DS, Soares MP (2012) Disease tolerance as a defense strategy. Science 335(6071):936–941
27. Soares MP, Gozzelino R, Weis S (2014) Tissue damage control in disease tolerance. Trends Immunol 35(10):483–494
28. Varela L, Martinez-Sanchez N, Gallego R et al (2012) Hypothalamic mTOR pathway mediates thyroid hormone-induced hyperphagia in hyperthyroidism. J Pathol 227(2):209–222
29. Kawa MP, Grymula K, Paczkowska E et al (2010) Clinical relevance of thyroid dysfunction in human haematopoiesis: biochemical and molecular studies. Eur J Endocrinol 162(2):295–305
30. Vitale G, Fatti LM, Prolo S et al (2010) Screening for hypothyroidism in older hospitalized patients with anemia: a new insight into an old disease. J Am Geriatr Soc 58(9): 1825–1827
31. Angelin-Duclos C, Domenget C, Kolbus A et al (2005) Thyroid hormone T3 acting through the thyroid hormone alpha receptor is necessary for implementation of erythropoiesis in the neonatal spleen environment in the mouse. Development 132(5):925–934

32. Zhang Y, Xue Y, Cao C et al (2017) Thyroid hormone regulates hematopoiesis via the TR-KLF9 axis. Blood 130(20):2161–2170
33. Kendrick TS, Payne CJ, Epis MR et al (2008) Erythroid defects in TRalpha-/- mice. Blood 111(6):3245–3248
34. Gianoukakis AG, Leigh MJ, Richards P et al. Chara (2009) Caracterization of the anaemia associated with Graves' disease. Clin Endocrinol 70(5):781–787
35. Linares M, Marin-Garcia P, Perez-Benavente S et al (2013) Brain-derived neurotrophic factor and the course of experimental cerebral malaria. Brain Res 1490:210–224

Chapter 7

Immunophenotyping of Hematopoietic Cells in the Spleen in Hypothyroid Mice

Diego Rodríguez Muñoz

Abstract

Hypothyroidism, which is characterized by insufficient production of thyroid hormones, and malaria, a mosquito-borne infectious disease caused by *Plasmodium* parasites, are significant global health challenges. Studying how these two conditions interact could provide valuable insights into their complex relationship and potential treatment options.

To induce hypothyroidism in the research, scientists used drugs to block the production of thyroid hormones. Then, they infected mice with *Plasmodium berghei* ANKA to mimic cerebral malaria infection. The spleen is essential in the body's immune response to malaria. It is involved in both innate and adaptive immunity, iron recycling, and the removal of old red blood cells or damaged cells infected with *Plasmodium*. Monitoring disease progression in male mice is crucial for early detection, and techniques like flow cytometry can help identify specific immune system populations within the spleen that are relevant to the research.

Key words Hypothyroidism, Spleen, Malaria, *Plasmodium berghei ANKA*, Flow cytometry, Leukocytes

1 Introduction

Malaria, also known as paludism, remains a devastating infection with significant health implications (85 endemic countries). As one of the most prevalent parasitic diseases worldwide, it exacts a heavy toll in terms of mortality, morbidity, and socioeconomic impact [1].

The causative agent of malaria is a unicellular protozoan from the genus *Plasmodium*. In humans, the infection spreads through dipteran vectors, primarily Anopheles, Aedes, or Culex mosquitoes [2]. These parasites enter the bloodstream via mosquito bites and undergo various stages of their life cycle within the human host, including replication inside liver cells and red blood cells [3].

Cerebral malaria (CM) represents the most lethal form of the disease [4]. It disproportionately affects children under the age of 5 [5]. Despite a decline in overall infection rates over the past

Constanza Contreras-Jurado (ed.), *Thyroid Hormones: Methods and Protocols*, Methods in Molecular Biology, vol. 2876,
https://doi.org/10.1007/978-1-0716-4252-8_7,

decade, the mortality rate from cerebral malaria remains alarmingly high approaching 10% in young children [6, 7]. The devastating effects of CM manifest at both the organ and molecular levels.

CM is characterized by disruption of the blood-brain barrier, leading to increased intracranial pressure and compression of cerebral blood vessels due to brain swelling [8]. This cascade results in altered consciousness, seizures, partial or total paralysis, and ultimately, coma and death [2, 4, 7, 9]. In children, *Plasmodium falciparum* directly lyses red blood cells, and the parasites have been implicated in deleterious effects on the gastrointestinal system and the brain [10–12].

Researchers commonly study malaria progression using the *Plasmodium berghei* ANKA (PbA) infection model in C57BL/6 J mice [13]. While this strain does not consistently develop cerebral malaria in all individuals, it provides valuable insights. The disease's outcome depends significantly on the intricate interplay between parasite and host [14, 15]. When healthy mice are inoculated with infected erythrocytes, they recapitulate pathological symptoms akin to cerebral malaria, including ataxia, weakness, hunching, seizures, paralysis, coma, and death within 5 to 7 days [16]. Remarkably, this process shares numerous similarities with the pathological mechanisms observed in human infections caused by *P. falciparum* [15, 17]. At the cellular level, elevated cytokine levels and increased expression of ICAM-1, VCAM-1, and P-selectin mirror effects are seen in human cases [14, 18].

The spleen plays a vital role in the immune response to malaria. It is responsible for the induction of innate and adaptive immune responses, the recycling of iron, and the elimination of senescent or damaged erythrocytes when they undergo the process of *Plasmodium* infection [19].

One of the most notable effects observed after *P. falciparum* infection in humans is splenomegaly [19]. This same reaction has been observed in the spleen of PbA-infected mice, which serves as an experimental model of the infectious process produced by *P. falciparum* in humans [20].

In malaria, both in rodent models and in humans, it has been described that the elimination function in the spleen of *Plasmodium*-infected or deformed erythrocytes is a determinant process in the development of the infection [21, 22]. It is also important to note the relevance of different splenic cell populations in host resistance to disease. Proinflammatory monocytes and dendritic cells are an important source of proinflammatory cytokines and reactive oxygen and nitrogen species that help generate a microenvironment conducive to *Plasmodium* destruction by phagocytic cells in both murine and human malaria [23, 24]. Macrophages, especially those in the red pulp of the spleen, are primarily responsible for the elimination of infected erythrocytes [19]. Monocytes, in turn, also play a key role in the course of the disease, being a major source of macrophage and specific dendritic cell generation [23].

Dendritic cells play a pivotal role in the activation of T cells through TLRs, such as TLR9, and the activation of natural killer (NK) cells through their capacity to produce IL-12 [25]. Upon activation, CD8+ T cells are capable of contributing to the elimination of infected erythrocytes by macrophages through the induction of phosphatidylserine expression on the surface of infected erythrocyte cells and the facilitation of their phagocytosis by the macrophage [26]. Activation of NKs by dendritic cells enables dendritic cells to start secreting IFN-γ, which promotes the differentiation of CD4+ T cells into Th1, a critical step in the acquisition of immunity against *Plasmodium* infections [27]. The ability of CD4+ T cells to condition phagocytic cells for *Plasmodium* clearance and to cooperate with B cells in the production of antimalarial antibodies has also been described [28].

The immune system serves as a vigilant defender against invading pathogens, while the host organism can also mitigate the adverse effects of infections on its fitness. An intriguing aspect of host defense is the capacity to tolerate the presence of pathogens—a unique approach that depends on the nature of the pathogen and complements other immune mechanisms [27, 28].

Recent studies have revealed an unexpected connection between hypothyroidism and tolerance to cerebral malaria. By modulating the host's metabolism to protect tissues from damage, hypothyroidism induces tolerance to infections, potentially enhancing survival. However, it also raises questions about the role of thyroid gland activity during malaria infection [8].

Thyroid hormones play critical roles in regulating key metabolic enzymes, both in peripheral tissues and the brain. For instance, T3 (triiodothyronine) controls energy consumption by modulating mTORC1 activity in the arcuate nucleus of the hypothalamus [29]. Evidence suggests that thyroid hormones also influence hematopoiesis and the immune response in both humans and animals. Thyroid receptor α has been implicated in the maturation of erythrocyte progenitors. Disruptions in thyroid balance significantly impact red blood cell development, leading to anemia a common symptom in patients with hypothyroidism [30–37].

This chapter serves a dual purpose. First, it explores how mice develop hypothyroidism and subsequently exhibit the pathology associated with cerebral malaria infection. Second, it delves into the identification of hematopoietic cells in the spleen using flow cytometry analysis a technique that enables rapid and simultaneous measurement of multiple parameters within mixed cell populations.

2 Materials

2.1 Generation of Hypothyroid Mice

1. C57BL/6 male mice (*see* **Note 1**).
2. Propylthiouracyl.
3. Potassium perchlorate.
4. Control diet.
5. Potassium iodide.

2.2 Generation of Cerebral Malaria Infections

1. *Plasmodium berghei ANKA*.
2. Para-amino benzoic acid.
3. EDTA microtubes.
4. Sterile or ethanol-cleaned surgical instruments.
5. Methanol.
6. Wright's solution.
7. Glass slides.
8. Microscope with x100 objective.
9. MiliQ water.

2.3 Isolation and Preparation of Mouse Splenic Hematopoietic Cells

1. Sterile or ethanol-cleaned dissection material.
2. 70% ethanol.
3. Mouse adult spleen dissociation kit.
4. 70-μm cell strainer.
5. ACK lysing buffer: 0.15 M ammonium chloride, 0.01 M potassium bicarbonate, and 0.1 mM disodium EDTA in distilled water. Adjust the pH to 7.2–7.4.
6. FACS buffer: PBS with 2% FBS and 5 Mm EDTA. Filtered.

2.4 Cell Surface Staining

All antibodies described herein are used at predetermined optimal concentrations (*see* **Note 2**).

1. FACS tubes: 12 × 75 mm polystyrene round bottom tubes.
2. Fluorochrome-labeled anti-mouse antibodies for flow cytometry against indicated cell surface markers: CD11b, CD11c, CD206a, MHCII, F4/80, SIGLEC H (CD169). Clone and fluorochromes are described in Table 1 (*see* **Note 3**).
3. Microbeads Perfect-Count Microspheres.

2.5 Equipment and Supplies

1. Centrifuge.
2. Cytometer FACSCanto II (Becton and Dickinson).
3. FlowJo software.

Table 1
Fluorochrome-labeled anti-mouse antibodies for flow cytometry

Antibodies specificity	Fluorochrome	Clone
CD11b	APC-Cy7	MI-70
CD11c	PerCP-Cy5.5	HL3
CD206a	BV 650	CO682C
SIGLEC H (CD169)	PE-Cy7	eBio440c
F4/80	BV 421	T45–2342
MHCII	Super Bright 600	M5/114.15.2

3 Methods

3.1 Generation of Hypothyroid Mice

1. Wean male mice 4–5 weeks after birth and provide them with an iodine-deficient diet containing 0.15% of the antithyroid drug propylthiouracil for four weeks, allowing them to eat as much as they want.
2. At this point, add potassium perchlorate to the drinking water at a concentration of 10 grams per liter. Change the drinking water daily.
3. For the control group of euthyroid mice, feed the animals with a control diet that is the same as the iodine-deficient diet but without propylthiouracil. Additionally, add potassium iodide at a concentration of 1.15 mg/kg to their diet. Do not change either diet throughout the experiment.

3.2 Generation of Cerebral Malaria Infections

1. After 4 weeks of hypothyroid treatment, inoculate the mice with *P. berghei* ANKA (PbA) by intraperitoneal injection of 1×10^6 infected red blood cells from donor mice, as detailed below.
2. Euthanize donor mice following a locally approved protocol (*see* **Note 1**).
3. Collect the blood into an EDTA tube via cardiac puncture and keep it at room temperature (RT).
4. Use one drop of blood to prepare a smear on a glass slide and let it dry at RT.
5. Fix the smear with methanol and allow it to dry at RT.
6. Stain the smear with Wright's solution and incubate for 3 min at RT.
7. Rinse the slide with Milli-Q water to completely remove the staining solution and let it dry at RT.

8. Examine the slide under a x100 objective and quantify the proportion of infected erythrocytes (*see* **Note 4**).
9. Calculate the required blood volume to inoculate each mouse with 1×10^6 infected red blood cells in 100 μl. Dilute the blood with sterile PBS 1X if necessary.
10. After inoculation, add 0.05% para-aminobenzoic acid to the drinking water to support the parasite life cycle. This time point is designated as day 0 post-infection.
11. To monitor parasite levels, collect a blood drop from the tail to prepare a smear daily, starting from days 2–3 post-infection. Follow **steps 4** to **8** of this section to determine parasite levels.
12. Observe and monitor all mice daily to assess the clinical stage of cerebral malaria. Use the following criteria to classify disease symptoms [37]:
    (0). No symptoms.
    (1). Presence of piloerection.
    (2). Moderate piloerection and mild ataxia.
    (3). Ataxia and paralysis.
    (4). Convulsions and coma.
13. Euthanize the mice (*see* **Note 1**) when they exhibit severe symptoms of the disease (stage 4).

3.3 Isolation and Preparation of Mouse Spleen Cells

This section details the dissection of mouse spleens and the subsequent isolation of spleen-infiltrating hematopoietic cells.

1. Euthanize the mice following an approved local protocol (*see* **Note 1**).
2. Disinfect the mouse skin with 70% ethanol using a spray bottle. Prepare 2 mL of buffer per mouse for spleen collection.
3. Carefully extract the spleens immediately after euthanasia, then mince and process them using the gentleMACS Dissociator with the Adult Spleen Dissociation Kit.
4. Filter the samples through a 70-μm cell strainer to eliminate tissue fragments.
5. Lyse erythrocytes by adding 400 μL of ACK lysing buffer for 2 min (*see* **Note 6**).
6. Add 2 mL of FACS buffer to the samples.
7. Centrifuge the cells at 300× *g* for 5 min at room temperature to wash them.
8. Resuspend the cells in an appropriate volume of FACS buffer.

3.4 Cell Surface Staining of Hematopoietic Spleen Cells

To determine the optimal concentration, each antibody is titrated before use. Antibody cocktails are prepared before staining the cells. The specified staining volumes can be adjusted based on cell numbers.

1. Allocate a small fraction of enriched cells into separate polypropylene tubes for FMO controls and samples.
2. Centrifuge at 300× *g* for 5 min at room temperature.
3. Resuspend the cells in 50 μL of FACS buffer.
4. Add an equal volume of antibody cocktail at twice the optimal concentration to achieve the final desired concentration of antibodies.
5. Incubate in the dark at room temperature for 20 min.
6. Add 500 μL of FACS buffer to wash the cells.
7. Centrifuge at 300 g for 5 min at room temperature.
8. Resuspend the cells in an appropriate volume for flow cytometry analysis.
9. One minute before cytometry analysis, add Perfect-Count Microspheres to measure the absolute number of cells.

3.5 Samples Acquisition

To minimize the risk of clogs, it is recommended to filter samples before conducting cytometry analysis. The flow cytometry data were collected using a FACSCanto II system and analyzed with FlowJo software.

3.5.1 Compensation Procedures

1. Utilize an aliquot of unstained cells as a negative control to set the appropriate gains for forward and side scatter, as well as to determine the optimal flow rate.
2. For each fluorochrome antibody, use a single-color positive control to adjust the flow cytometer's voltage settings and mix compensation as described below.
3. Prepare single-stained compensation controls by aliquoting 100 μL of cells (0.5×10^6 cells).
4. Add the antibody in the correct volume to achieve the final optimal concentration. Create as many single-color controls as there are antibodies in the cocktail.
5. Follow **steps 5** to **8** outlined in Subheading 3.4.
6. Acquire all single-stained controls, including the negative control, and calculate compensation values across all relevant detectors as per software guidelines.
7. For sample acquisition, gather enough events to ensure statistically significant data.

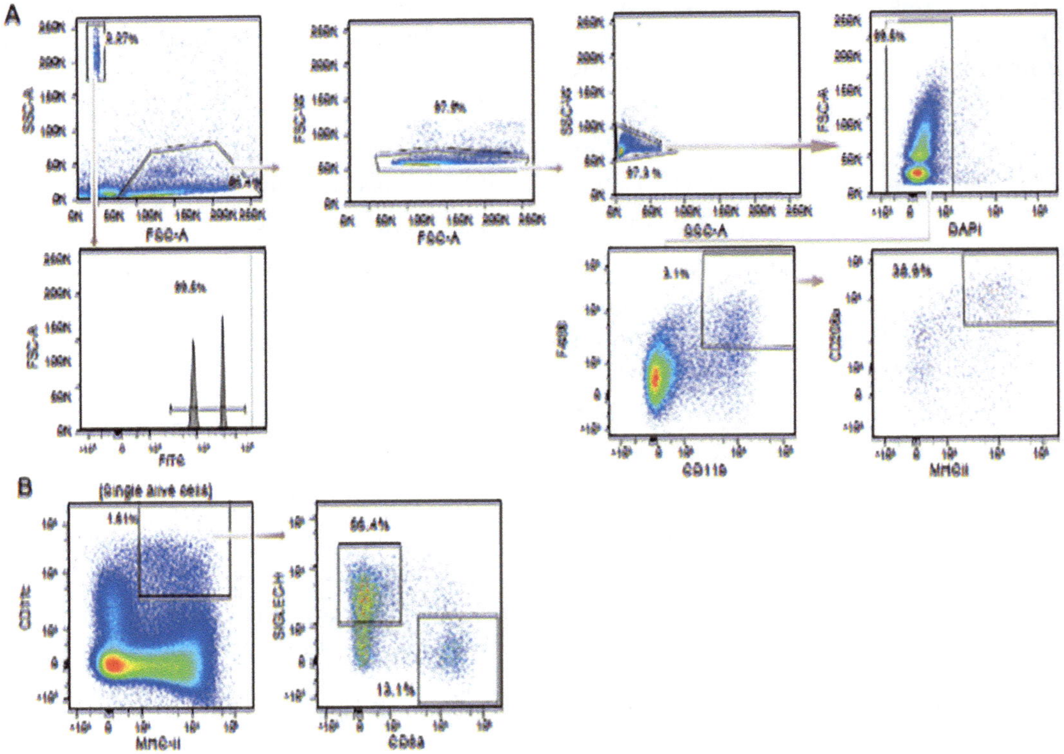

Fig. 1 (**a**) Individual live splenic cells were sorted by F4/80 and CD11b expression. Mono-DCs were identified by an additional strategy of F4/80 + CD11b + with CD206a against MHCII. (**b**) Individual live splenic cells were sorted by CD11c and MHCII expression. CD11c+/MHCII+ cells were then divided and sorted into two subgroups based on their positive expression of Siglec H (pDCs) or CD8a (cDC1s)

3.5.2 Gate Setting Strategies

This section outlines a method for establishing gates to identify various immune cell populations of interest. The discussion references plots shown in Fig. 1, as described in a prior study [8]. Notably, many parameters are displayed using a biexponential axis, which employs an alternative scaling at the lower end to accommodate negative values and avoid events clustering along the axes, thus enhancing data visualization. The initial three steps focus on excluding most cellular doublets: Plotting SSC-area against FSC-area excludes smaller or larger particles such as unwanted cells and debris.

1. Plotting FCS-weight against FCS-area allows for the exclusion cellular doublets.
2. Plotting SSC-weight against SSC-area allows for the exclusion cellular doublets.
3. Plotting FCS-area against DAPI for select viable cells to continue.

4. The approach is structured into two parts. First, to identify monocyte-derived dendritic cells (Mo-DCs) and second to detect dendritic cells (pDCs and cDC1s).
   4.1. Single alive splenic cells were plotted for F4/80 and CD11b. Double positives were plotted again for CD206a and MHCII, being Mo-DCs identified such CD206a + MHCII+.
   4.2. Plotting MHCII against CD11c to select the double positive cell population. Continue plotting Siglec H against CD8a to select the single positive cell population. pDCs are identified such Siglec H+ CD8a + (cDC1s).

4 Notes

1. All experimental procedures involving animals must be approved by local ethics committees and comply with national regulations.
2. Although the optimal antibody concentration per million cells is provided by the manufacturer, it is advisable to adjust this concentration based on the specific cell suspension and cytometer used. Variations in laser intensity between cytometers and different antibody lots necessitate titration of each antibody prior to experimentation.
3. The antibodies used in this study are commercially available from multiple suppliers, with the majority sourced from BD Biosciences (San Jose, CA), eBioscience (San Diego, CA), and BioLegend (San Diego, CA).
4. Any image quantification software can be used; in this study, Fiji software was employed for sample quantification.
5. Centrifugation without braking is crucial to prevent disruption of the distinct phases formed during the process.
6. This step can be omitted if erythrocytes are not present.

References

1. WHO (2023) World malaria report 2023. World Health Organizartion. https://www.who.int/publications/i/item/9789240086173
2. Boyle MJ, Engwerda CR, Jagannathan P (2024) The impact of plasmodium-driven immunoregulatory networks on immunity to malaria. Nat Rev Immunol 9:637–653
3. Schwarzer E, Skorokhod O (2024) Post-translational modifications of proteins of malaria parasites during the life cycle. Int J Mol Sci 25(11):6145
4. Sharma I, Kataria P, Das J (2024) Cerebral malaria pathogenesis: dissecting the role of $CD4^+$ and $CD8^+$ T-cells as major effectors in disease pathology. Int Rev Immunol 15:1–18. https://doi.org/10.1080/08830185.2024.2336539. Epub ahead of print
5. Kwansa-Bentum H, Aninagyei E, Adedia D et al (2024) Elevation of free triiodothyronine (fT3) levels by plasmodium falciparum

independent of thyroid stimulating hormone (TSH) in children with uncomplicated malaria. J Clin Lab Anal 38(3):e25013
6. Cowman AF, Healer J, Marapana D et al (2016) Malaria: biology and disease. Cell 167(3):610–624
7. Rodriguez-Muñoz D, Sánchez Á, Pérez-Benavente S et al (2022) Hypothyroidism confers tolerance to cerebral malaria. Sci Adv 8(14): eabj7110
8. Nkumama IN, O'Meara WP, Osier FHA (2017) Changes in malaria epidemiology in Africa and new challenges for elimination. Trends Parasitol 33(2):128–140
9. Aninagyei E, Doku ET, Adu P et al (2018) Storage related haematological and biochemical changes in plasmodium falciparum infected and sickle cell trait donor blood. BMC Hematol 18(1):30
10. Agrawal D, Teach SJ (2006) Evaluation and management of a child with suspected malaria. Pediatr Emerg Care 22(2):127–133
11. Crawley J, Chu C, Mtove G et al (2010) Malaria in children. Lancet 375(9724): 1468–1481
12. Craig AG, Grau GE, Janse C et al (2012) The role of animal models for research on severe malaria. PLoS Pathog 8(2):e1002401
13. Lou J, Lucas R, Grau GE (2001) Pathogenesis of cerebral malaria: recent experimental data and possible applications for humans. Clin Microbiol Rev 14(4):810–820
14. Idro R, Jenkins NE, Newton CR (2005) Pathogenesis, clinical features, and neurological outcome of cerebral malaria. Lancet Neurol 4(12):827–840
15. Nishanth G, Schluter D (2019) Blood-brain barrier in cerebral malaria: pathogenesis and therapeutic intervention. Trends Parasitol 35(7):516–528
16. de Souza JB, Hafalla JC, Riley EM et al (2010) Cerebral malaria: why experimental murine models are required to understand the pathogenesis of disease. Parasitology 137(5): 755–772
17. Carvalho LJ (2010) Murine cerebral malaria: how far from human cerebral malaria? Trends Parasitol 26(6):271–272
18. Del Portillo HA, Ferrer M, Brugat T, Martin-Jaular L, Langhorne J, Lacerda MV (2012) The role of the spleen in malaria. Cell Microbiol 14(3):343–355
19. Lakkavaram A, Lundie RJ, Do H, Ward AC, de Koning-Ward TF (2020) Acute Plasmodiu berghei mouse infection elicits perturbed erythropoiesis with features that overlap with anemia of chronic disease. Front Microbiol 11: 702
20. Engwerda CR, Beattie L, Amante FH (2005) The importance of the spleen in malaria. Trends Parasitol 21(2):75–80
21. Mebius RE, Kraal G (2005) Structure and function of the spleen. Nat Rev Immunol 5(8):606–616
22. Hirako IC, Ataide MA, Faustino L, Assis PA, Sorensen EW, Ueta H et al (2016) Splenic differentiation and emergence of CCR5(+) CXCL9(+) CXCL10(+) monocyte-derived dendritic cells in the brain during cerebral malaria. Nat Commun 7:13277
23. Gazzinelli RT, Kalantari P, Fitzgerald KA, Golenbock DT (2014) Innate sensing of malaria parasites. Nat Rev Immunol 14(11): 744–757
24. Gowda NM, Wu X, Gowda DC (2012) TLR9 and MyD88 are crucial for the development of protective immunity to malaria. J Immunol 188(10):5073–5085
25. Imai T, Ishida H, Suzue K, Taniguchi T, Okada H, Shimokawa C et al (2015) Cytotoxic activities of CD8(+) T cells collaborate with macrophages to protect against blood-stage murine malaria. elife:4
26. Artavanis-Tsakonas K, Riley EM (2002) Innate immune response to malaria: rapid induction of IFN-gamma from human NK cells by live plasmodium falciparum-infected erythrocytes. J Immunol 169(6):2956–2963
27. Kumar R, Loughland JR, Ng SS, Boyle MJ, Engwerda CR (2020) The regulation of CD4 (+) T cells during malaria. Immunol Rev 293(1):70–87
28. Medzhitov R, Schneider DS, Soares MP (2012) Disease tolerance as a defense strategy. Science 335(6071):936–941
29. Soares MP, Gozzelino R, Weis S (2014) Tissue damage control in disease tolerance. Trends Immunol 35(10):483–494
30. Varela L, Martinez-Sanchez N, Gallego R et al (2012) Hypothalamic mTOR pathway mediates thyroid hormone-induced hyperphagia in hyperthyroidism. J Pathol 227(2):209–222
31. Kawa MP, Grymula K, Paczkowska E et al (2010) Clinical relevance of thyroid dysfunction in human haematopoiesis: biochemical and molecular studies. Eur J Endocrinol 162(2):295–305
32. Vitale G, Fatti LM, Prolo S et al (2010) Screening for hypothyroidism in older hospitalized patients with anemia: a new insight into an

old disease. J Am Geriatr Soc 58(9): 1825–1827

33. Angelin-Duclos C, Domenget C, Kolbus A et al (2005) Thyroid hormone T3 acting through the thyroid hormone alpha receptor is necessary for implementation of erythropoiesis in the neonatal spleen environment in the mouse. Development 132(5):925–934
34. Zhang Y, Xue Y, Cao C et al (2017) Thyroid hormone regulates hematopoiesis via the TR-KLF9 axis. Blood 130(20):2161–2170
35. Kendrick TS, Payne CJ, Epis MR et al (2008) Erythroid defects in TRalpha-/- mice. Blood 111(6):3245–3248
36. Gianoukakis AG, Leigh MJ, Richards P et al (2009) Characterization of the anaemia associated with Graves' disease. Clin Endocrinol 70(5):781–787
37. Linares M, Marin-Garcia P, Perez-Benavente S et al (2013) Brain-derived neurotrophic factor and the course of experimental cerebral malaria. Brain Res 1490:210–224

Chapter 8

Mouse Embryonic Fibroblasts Reprogramming to Induced Pluripotent Stem Cells by T3

Ana Montero-Pedrazuela and Constanza Contreras-Jurado

Abstract

Somatic cells can be transformed into induced pluripotent stem cells (iPSCs) using a technique called reprogramming. This process involves introducing Yamanaka factors (Oct4, Sox2, Klf4, and c-Myc) to the cells through retroviral supernatants. This chapter outlines a protocol for reprogramming mouse embryonic fibroblasts (MEFs) using the hormone triiodo-L-thyronine (T3) to enhance the generation of iPSCs. It also describes how to analyze these iPSCs by colony staining for alkaline phosphatase activity, a standard marker for identifying pluripotent embryonic stem cells. To further study iPSCs, individual colonies must be selected and expanded, and pluripotency is examined by analyzing gene expression profiles using quantitative real-time PCR to measure the endogenous expression of pluripotency genes. Integrating T3 into reprogramming methods may significantly improve the production of functional iPSCs. This advancement could open new avenues for research in cell plasticity, disease modeling, and regenerative therapies.

Key words Thyroid hormone, Mouse embryonic fibroblasts, Reprogramming, Pluripotency, Quantitative real-time PCR, Yamanaka factors, iPSC colony formation

1 Introduction

Pluripotency refers to the ability of individual cells to initiate all lineages of the mature organism in a flexible manner, directed by signals in the embryo or cell culture environment. It is a crucial aspect of mammalian development and embryonic stem cell (ES) biology. Reprogramming to induced pluripotent stem cells (iPSCs) can be generated from somatic mouse [1] and human [2] cells by the so-called Yamanaka factors, namely Oct4, Sox2, Klf4, and c-Myc (commonly abbreviated as OSKM). The discovery of iPSC reprogramming offers an alternative to embryonic stem cells (ESCs), with fewer ethical concerns since they are not derived from fertilized pre-implantation embryos. Personalized iPSCs offer potential for disease cure exploration, drug screening, and custom regenerative medicine. During chromatin reprogramming, pioneer transcription factors directly bind to condensed chromatin

Constanza Contreras-Jurado (ed.), *Thyroid Hormones: Methods and Protocols*, Methods in Molecular Biology, vol. 2876, https://doi.org/10.1007/978-1-0716-4252-8_8,

[3]. This initiates a series of chromatin remodeling events, including modification of histone H3 lysine 4 and 27. These events lead to opening of chromatin during cell division, creating an open chromatin environment that allows other transcription factors to be readily accessible [4]. The rapid proliferation of iPSCs, characterized by short G1- and G2-gap phases, is accompanied by self-renewal and pluripotency, which are signatures of cell cycle reprogramming [5]. At an early stage of reprogramming, mesenchymal-to-epithelial transition (MET) occurs. This transition is characterized by the upregulation of epithelial genes, including E-Cadherin, Cdh1, and Epcam, and is a critical event toward deriving iPSCs from mouse embryonic fibroblasts (MEFs) [6].

Several nuclear receptors have been identified to play a crucial role in obtaining and maintaining pluripotency in human and mouse cells [7]. Certain coactivators and corepressors that interact with nuclear receptors have also the ability to influence the process of somatic cell reprogramming [8, 9]. Thyroid hormones are essential for metabolism, development, and growth. Triiodo-L-thyronine (T3) is the most active form of thyroid hormones and plays a significant role in regulating gene expression by binding to nuclear receptors (TRs) [10]. These receptors act as hormone-dependent transcription factors by binding to DNA motifs located in the regulatory regions of target genes. They also recruit coregulators, such as coactivators and corepressors, which alter chromatin structure [11].

In our lab, we have demonstrated that T3 enhances the reprogramming of somatic MEFs into pluripotent stem cells [12]. To study the effect of T3 on reprogramming, MEFs were transduced with the Yamanaka factors and treated with medium supplemented with T3. The generation of iPSCs was assessed by observing the formation of alkaline phosphatase-positive colonies, analyzing pluripotency gene expression, and assessing embryoid body formation. In this chapter, we will describe how to enhance reprogramming of MEFs transduced with the Yamanaka factors to iPSCs by T3 treatment. The method we have developed involves assessing the formation of alkaline phosphatase-positive colonies and evaluating the gene expression profile to detect the acquisition of pluripotency by using quantitative real-time PCR to detect the endogenous expression of Sox2, Rex1, Oct4, and Nanog genes.

2 Materials

Please ensure that you use ultrapure water for preparing all solutions. This can be achieved by purifying deionized water to attain a sensitivity of 18 MΩ-cm at 25 °C. Use analytical grade reagents for all preparations. Also, make sure to follow all waste disposal regulations when disposing of waste materials.

2.1 Primary Cultures and Cells Lines

1. MEFs from wild-type C57BL/6 mice obtained from embryos at E13.5 and cultured in Standard Medium.
2. HEK293T cell line for transfection experiments (cultured in Standard Medium).
3. Murine ES cell line R1 (ATTC, SCRC-1011) as a control for stemness and cultured in iPSCs Medium.

2.2 Cell Culture Media and Reagents

1. Dissecting reagents: sterile PBS + antibiotic, 95% ethanol, trypsin.
2. Trypan Blue, DMSO.
3. Standard Medium: DMEM supplemented with 10% of fetal bovine serum (FBS) and penicillin-streptomycin.
4. iPSC Medium: high glucose DMEM (with pyruvate and stable glutamine), supplemented with 15% knockout serum replacement (KSR, Gibco, or similar xeno-free formulated serum substitute; *see* **Note 1**), leukemia inhibitory factor (LIF; 1000 U/mL), non-essential amino acids, penicillin-streptomycin, and 50 mM β-mercaptoethanol.
5. Triiodothyronine (T3) stock solution: 1 mM T3 in 0.04 M NaOH.
6. T3 working solution: dilute stock solution to 1 μM with DMEM. Store frozen in single-use aliquots. Final concentration in culture: 10 nM.

2.3 Viral Transduction, MEFs Infection and Reprogramming

1. Exogenous Yamanaka factors: retroviral plasmids pMXs-Oct4, pMXs-Sox2, pMXs-Klf4, or pMXs-cMyc (Addgene).
2. pCL-Eco packaging plasmid (Addgene).
3. pBABE-PURO-GFP plasmid (Addgene) or similar for transfection efficiency confirmation.
4. Fugene 6 (Promega) or similar transfection reagent.
5. 8 mg/mL Polybrene (1000x).

2.4 Colony Quantification

1. 4% paraformaldehyde in phosphate buffer pH 7 (PFA).
2. Alkaline phosphatase (AP) activity: colorimetric AP kit (NBT/BCIP-based).

2.5 Quantitative Real-Time PCR

1. Total RNA extraction kit.
2. DNase I.
3. Reverse transcription system.
4. Power SYBR Green PCR Master Mix (Applied Biosystems) or similar product.

5. Oligonucleotides for SYBR Green-based qRT-PCR: Forward primer (5′-3′)/Reverse primer (5′-3′)

   mNanog NM_028016

   CAAGGGTCTGCTACTGAGATGCTCTG/TTTTGTTTGG GACTGGTAGAAGAATCAG

   mOct4 NM_013633

   GTTGGAGAAGGTGGAACCAA /CCAAGGTGATCCTCT TCTGC

   mRex1 NM_009556

   ACGAGTGGCAGTTTCTTCTTGGGA/TATGACTCACTT CCAGGGGGCACT

   mSox2 NM_011443

   AAAGGAGAGAAGTTTGGAGCCCGA/GGGCGAAGTGC AATTGGGATGAAA

   mFbn2 NM_010181

   ACGCAAATCAATTCAGCAGTGT/GGTTGTCCACAGTA AGTTCCGA

   mTRα1 NM_178060

   GGCTGTGCTGCTAATGTCAA/CGGAGGTCAGTCACC TTCAT

   mTRβ NM_001113417

   AGACAAAGTCACCCGCAACC/CTAGCCTCTTGCTGT CATCC

   Reference gene: mGusb NM_010368

   GAGGATCAACAGTGCCCATT/CAGCCTCAAAGGGGA GGT.

2.6 Plastic Material

1. Sterile 6-well, 24-well, and V-bottomed 96-well plates.
2. Sterile 100-mm and 60-mm dishes.
3. Sterile T75 and T150 flasks.
4. 15-mL sterile conical tubes.

2.7 Instruments

1. Dissecting tools: Scissors, forceps, scalpel (at least 2 of each).
2. Tissue culture hood.
3. Cell culture humidified incubator (37 °C, 5% CO_2).
4. Centrifuge.
5. Dissecting microscope.
6. qRT-PCR apparatus and software for analysis.

3 Methods

3.1 Harvesting and Culturing MEFs

1. Place clean forceps and scissors in one beaker with ethanol and prepare another with PBS to wash instruments between embryos. Distribute sterile PBS + antibiotic into 100-mm (1 per uterus) and 60-mm dishes (2–3 per embryo).
2. Euthanize the pregnant female at E13.5 post-conception following standard protocols (avoid CO_2 inhalation, as this causes acidosis).
3. Saturate fur with ethanol and make a vertical incision in the skin. Pin the skin and body wall to the side.
4. Open the body wall using fresh forceps and scissors, exposing the intact uterus.
5. Extract the uterus with forceps, ensuring it doesn't contact non-sterile surfaces. Cut the cervix and blood vessels, transferring the uterus to the 100-mm dish with PBS + antibiotic.
6. Work in a tissue culture hood. Cut the uterus between each embryo, placing each uterine section with a single embryo in a fresh 60-mm dish with PBS + antibiotic. Extract each embryo through the cut portion of the uterus and wash with PBS to remove maternal blood cells.
7. Transfer the embryo to a fresh 60-mm dish with sterile PBS, remove membranes and the umbilical cord. Incise the abdominal wall vertically, removing hematopoietic tissue and tubular intestine (pink colored).
8. Dissect away bulk CNS tissue by removing the head above the oral cavity (*see* **Note 2**).
9. Place the embryo body in a 60-mm dish with 2 mL of 1x trypsin, mince ~30 times in different angles using a scalpel, and incubate at 37 °C for 20 min.
10. Collect cells and tissue bits and reduce the size of bigger tissue bits by pipetting up and down ~15 times into a 15 mL conical tube. Incubate at 37 °C for another 20 min.
11. Add 15 mL of Standard Medium, resuspend the cell pellet and transfer to one T75 flask. Incubate at 37 °C in 5% CO_2, marking as "passage 0" with the date and embryo ID.
12. Change the medium the following day.
13. Passage primary MEFs when cells are nearly confluent (usually 1–3 days), splitting by two. These will be "passage 1" MEFs.
14. When confluent again (2–3 days), trypsinize, and determine viable cell count with a hemocytometer and Trypan Blue stain using an aliquot of the cell suspension.

15. These will be "passage 2" MEFs and could be directly used for reprogramming or frozen in aliquots of 3–4 × 10^6 cells in 90% FCS 10% DMSO until further use (*see* **Note 3**).

3.2 Culturing the Packaging Cell Line for Viral Production (Fugene 6 Method) and Infecting MEFs

Retroviral supernatants are generated using HEK-293 T cells transfected with the ecotropic packaging plasmid pCL-Eco alongside each one of the retroviral constructs: pMXs-Klf4, pMXs-Sox2, pMXs-Oct4, or pMXs-cMyc (*see* **Note 4**). Maintaining HEK293T cells in early passage and optimal condition is crucial for achieving high-yield retrovirus production. For this protocol, retrovirus production and MEFs infection are coupled (*see* **Note 5**).

1. Day 1: Inoculate 5 × 10^6 HEK293T cells per 100-mm dish with 10 mL of Standard Medium (*see* **Note 6**). Separate transfections will be needed for each Yamanaka factor, and for the positive control of transfection (GFP vector). The number of dishes will be related to the number of cells to be subsequently infected. Incubate the cells for around 2 h before performing the transfection.
2. Prepare the transfection for each factor to be used as follows (recipe for one 100-mm dish, scale when needed): in one polystyrene tube add DMEM (to a final volume of 600 μL) and 24 μL of FuGene 6 (*see* **Note 7**). Mix and incubate for 5 min. Add 4 μg of one pMX plasmid (either Sox2, or Oct4, or Klf4, or cMyc) and 4 μg of pCL-Eco plasmid, mix and incubate for 15 min at room temperature (*see* **Note 8**). Add the complete mix to one 100-mm dish, drop-by-drop, distributing the drops throughout the plate. Gently spin the dish to allow an even distribution of the complex before returning it to the incubator (37 °C, 5% CO_2).
3. If frozen MEFs are being infected, thaw the vial(s) in a 100-mm dish on the same day of HEK293T transfection (use only MEFs from early passages, *see* **Note 3**). If fresh MEFs are being infected, ensure they will be ready for the infection. Otherwise, frozen viral supernatant could be used (*see* **Note 5**).
4. Day 2: Replace the Standard Medium for the HEK293T cells.
5. Seed the MEFs at 1 × 10^5 per 6 well plate with 3 mL of Standard Medium. The initial culture density should have to be determined based on the rate of cell proliferation (due to the cells status, impact of mutations on proliferation if mutant-derived MEFs are used, or to the culture conditions) and on the objective of the assay (time of culture, addition of treatments, expected differences between conditions, ...).
6. Day 3: In the morning, collect the media from each individual transfection, pooling media for the same vectors from different dishes (if required) in the same tube. Replace the Standard Medium of the HEK293T plates before returning to the incubator.

7. Spin the collected media at 300 G for 5 min to eliminate floating cells and debris. Filter the media through 0.45 mm filters to avoid impurities.
8. Mix media from each transfection at equal proportions (e.g., 1:1:1:1 for Oct4:Klf4:Sox2:Myc infection) and add polybrene to achieve a final concentration of 8 μg/mL.
9. Remove media from the MEFs cultures and add 3 mL of the pooled retroviral media supplemented with polybrene to each 6-well plate containing the recipient MEFs.
10. Repeat **steps 7–10** every 10 to 12 h to achieve a total of 3 or 4 rounds of infection (*see* **Note 9**).
11. To quantify retroviral transduction efficiency, do parallel infections containing all the retroviruses used for reprogramming plus a GFP retroviral supernatant (pBABEpuro-GFP plasmid). Check the efficiency either by fluorescent microscopy or by flow cytometry at day 3 post-infection. Typical results are in the range of 40–60% for wild-type MEFs.
12. Eliminate the retroviral cultures and media following the local guidelines for viral inactivation and disposal.

3.3 MEFs Reprogramming

1. The day after completing the infection, replace the media with iPSC Medium. Cultures are sustained without drug selection, with daily medium changes (*see* **Note 10**).
2. For hormonal treatment add T3 to each well (to a final concentration of 10 nM) or equal volume of vehicle (*see* **Note 11**).
3. Seven to 10 days after infection, iPSC colonies will start to be apparent.
4. From day 10 to day 14, colonies with ES-like morphology became clearly visible and can be scored after alkaline phosphatase (AP) staining.
5. Use Murine ES cell line Rl (or similar) as a parallel control for stemness, cultured with iPSC medium.

3.4 Colony Staining and Quantification

AP expression can be easily measured by the ability of this enzyme to cause a colorimetric reagent to precipitate at the bottom of the culturing well (Fig. 1).

1. Discard the culture medium and perform a single wash with 1x PBS.
2. Fix the colonies for 5 min at room temperature using 4% paraformaldehyde in phosphate buffer pH 7 (PFA).
3. Remove the PFA and perform another wash with 1x PBS.
4. Prepare the alkaline phosphatase substrate NBT/BCIP following the manufacturer's instructions.

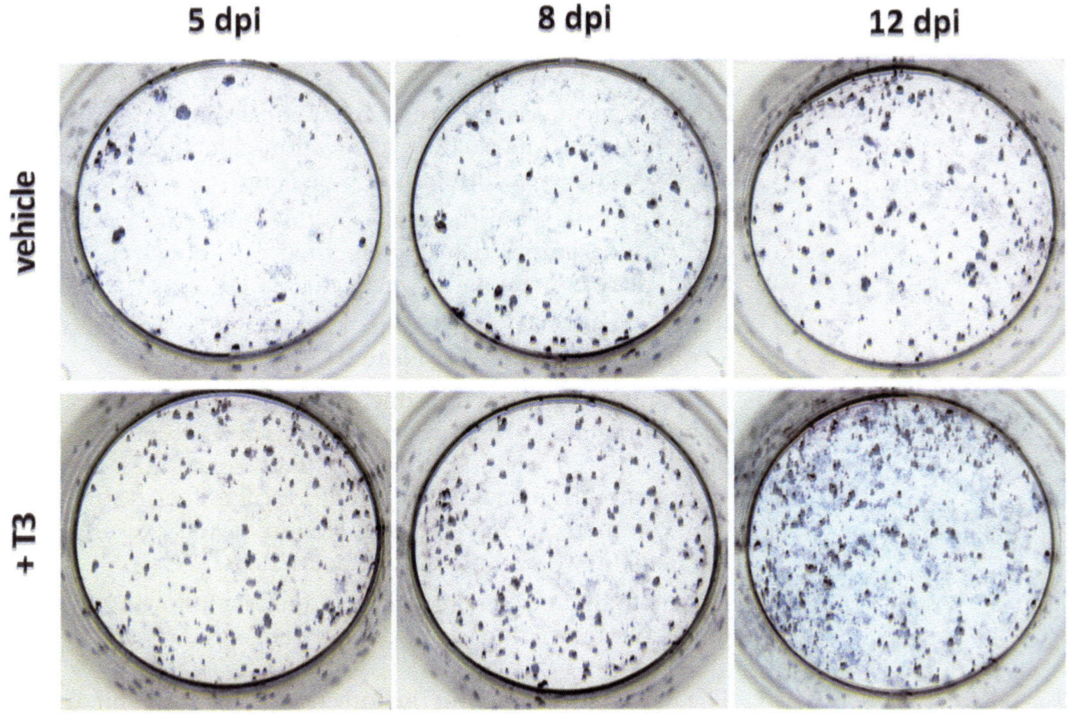

Fig. 1 Identification of iPSCs colony formation by alkaline phosphatase staining. Wells containing MEFs treated with iPSC reprogramming medium plus vehicle (upper panels) or plus T3 (lower panels) are shown, at different days post-infection (dpi) with the Yamanaka factors. Note the increasing number of colonies detected over the dpi, and the augmented number of colonies with T3 treatment in comparison to vehicle-treated wells of the same dpi

5. Incubate with the solution for 20 min to 1 h, check for background staining.
6. At the end of the incubation, perform a final wash with 1x PBS to halt the reaction.
7. Count the total number of AP+ colonies in each well. Well-defined criteria such as minimum size, intensity of marking, and shape (rounded) must be established for quantification (*see* **Note 12**). For comparison, it is advisable to run parallel wells with non-infected MEFs and cultures of an ES cell line with iPSC Medium. It is also recommended to perform the quantification at different time points and to use technical and biological replicates.

3.5 iPSC Colony Expansion

To further study iPSC colonies, select and expand individual colonies on MEF feeders using iPSC Medium.

1. Prior to picking the colonies, prepare the required number of gelatin-coated 24-well plates with inactivated MEFs (as feeders) as explained elsewhere [13].

2. Feed the colonies on the plates 1–2 h before picking with iPSC Medium.
3. Add 50 μL of trypsin 1x to the wells of a V-bottomed 96-well plate.
4. Wash the dish containing the colonies to pick with PBS, and add 1 mL of PBS to the dish (*see* **Note 13**).
5. Select well-defined and isolated colonies. Using a P20 pipette set on 20 μL and a dissecting microscope, pick one colony by detaching it from the surrounding layer of cells with the pipette tip and transfer to a 96-well with trypsin by pipetting up and down (*see* **Note 14**). Check for the colony inside the well, as it tends to stick to the plastic tip.
6. Incubate at 37 °C for 15 min and pipet up and down. Check under a dissecting microscope to confirm that the cells have been disaggregated.
7. Transfer the cells to the 24-well plate containing feeder cells and 450 μL of iPSC Medium. Add T3 10nM or vehicle to the corresponding wells.
8. Change the medium the next day to remove any remaining trypsin. Grow the cells until they form clearly visible colonies or nearly reach confluency, feeding them daily with iPSC Medium (and the appropriate treatments). The cells will be ready to expand in 5–7 days. If the colonies become too large, they must be passed through to prevent differentiation.
9. To expand the clones, prepare the required number of gelatin-coated 6-well plates with inactivated MEFs (feeders).
10. Once the colonies have reached the desired size, remove the medium, wash with 0.5 mL of PBS, remove all the PBS, and add 0.1 mL of trypsin. Incubate the cells at 37 °C for 10–15 min and pipette up and down. Check for disaggregation under a microscope.
11. Add 0.4 mL of iPSC Medium and suspend the cells by pipetting up and down to achieve a single-cell suspension.
12. Transfer the cell suspension to a well of a 6-well plate containing feeder cells and add 1.5 mL of iPSC Medium.
13. Feed the cells daily with iPSC Medium until they form clearly visible colonies (typically around a week). Collect the colonies for RNA extraction or prepare a frozen stock of the colonies until further use using iPSC freezing medium (20% DMSO 40% iPSC Medium 40% FBS).

3.6 Quantitative Real-Time PCR

The aim of this analysis is to study pluripotency gene expression (as *Nanog*, *Oct4*, *Rex1*, or *Sox2*), as opposed to a MEFs marker (*Fbn2*) (Fig. 2) or other genes of interest (here, *TRα1* and *TRβ1*) in iPSC after reprogramming and/or in the expanded colonies treated

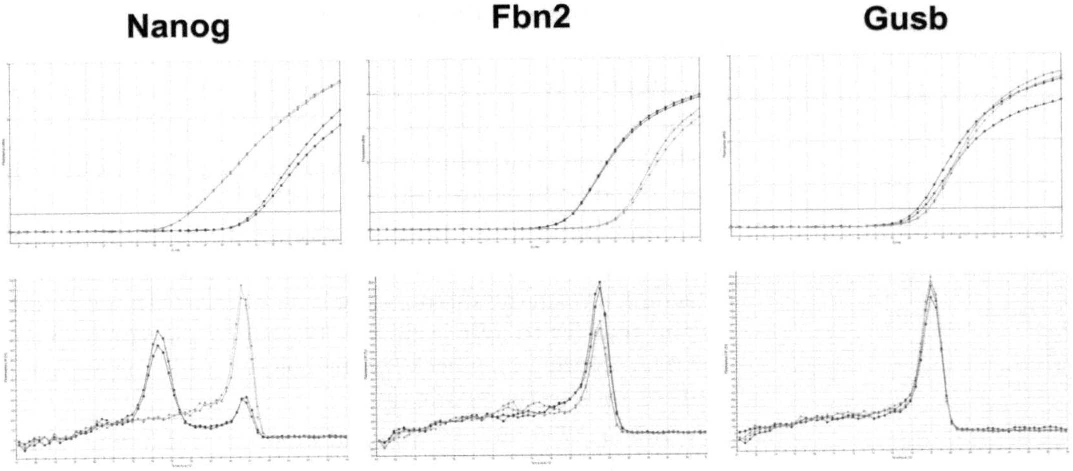

Fig. 2 Comparison of gene expression by qRT-PCR in reprogrammed vs non-reprogrammed MEFs. Upper panels show amplification plots and lower panels dissociation curves for the pluripotency gene Nanog, the MEF-specific gene Fbn1, and the reference gene Gusb (images captured from the MxPro analysis software for the Mx3005P Real-Time PCR System, Stratagene). Observe the clear separation of 2 populations: green and gray curves correspond to reprogramming MEFs (high expression of specific *Nanog*, lower expression of *Fbn1*) and blue and red curves correspond to non-reprogrammed MEFs (unspecific melting curve for *Nanog*, high expression of *Fbn1*). The four samples show a similar expression for the reference gene *Gusb*. Even in the presence of contaminating non-reprogrammed MEFs in the reprogrammed sample, the expression of *Fbn1* is significantly lower compared to non-reprogrammed MEFs. For the sake of simplicity, only 4 samples have been included in this example

or not with T3. Collecting the entire well content will include non-reprogrammed or feeder MEFs as contaminants (this should be acknowledged when analyzing the results). For expanded colonies growing on top of a feeder layer, an additional step for reducing contamination of feeder cells could be done: trypsinize the well and seed in an equal size gelatin-coated well. Incubate for 30–45 min and collect the supernatant containing mainly iPSC (feeder cells will attach to the well more quickly).

1. Sample Preparation: Clean the surfaces of any materials that will be used in the extraction of RNA with an RNase degradation reagent and always wear gloves to handle reagents and materials. Collect cells or colonies into a 15-mL tube (1 mL of trypsin +4 mL PBS per 6-well). Centrifuge at 300G for 5 min and discard the medium. Store cell pellets at −80 °C until RNA extraction. Due to the low cell number per well, we strongly recommend using a column-based RNA extraction kit to reach optimum yield and purity. Other RNA extraction methods such as TRIzol reagent could also be used, as long as high-quality enough RNA is obtained (collect the cells accordingly). An additional step of DNase I treatment is also advisable for

subsequent qRT-PCR using SYBR green. If TaqMan probes are used this treatment could be skipped as it may be not necessary.

2. Quantify the yield of total RNA spectrophotometrically at 260 nm, where 1 absorbance unit (A260) corresponds to 40 μg of single-stranded RNA per mL. Evaluate purity through spectrophotometry by assessing the relative absorbances at 230, 260, and 280 nm (A260/A280 and A260/A230). It is also advisable to determine RNA integrity (RIN) using an Agilent Bioanalyzer (Agilent, Santa Clara, CA) (*see* **Note 15**).
3. Carry out the reverse transcription to obtain a first-strand cDNA by means of a kit, in accordance with the manufacturer's protocol. Use 1 μg of RNA to obtain 1 μg of cDNA employing Oligo-dT primers. The obtained cDNA can be used in qPCR immediately or stored at −20 °C for later use.
4. Use a real-time PCR system to conduct the assay. Use technical triplicates to analyze mRNA levels using the forward and reverse primers specific for the target gene transcripts and the reference gene described in Subheading 2.5, along with an appropriate master mix (such as Power SYBR Green PCR Master Mix). Program the instrument with the appropriate settings for the primers used (in particular, annealing temperature). If using SYRB green, it is recommended to add a final step with a dissociation curve to analyze the specificity of the amplified sequence (Fig. 2).
5. Use the data obtained from the apparatus (number of cycles or Ct, with normalization based on a reference dye) to perform the calculations using the $2^{(-\Delta\Delta Ct)}$ method with *Gusb* mRNA serving as a reference gene [14]. Calculate the average of the replicates, normalize the results of the gene of interest to the results obtained from the reference gene (*Gusb*), and normalize the results of the target gene for T3 treated samples to those obtained from control samples. Perform statistical analysis using the final data.

4 Notes

1. This protocol is based on using a formulated serum replacement instead of FBS due to a better control of media composition. FBS would produce better results in terms of reprogramming efficiency, but more variable between product stocks. Be careful when using a formulated serum replacement, as formulas are usually proprietary and specific composition is not available. You should contact the manufacturer to be certain if the serum replacement contains any metabolite of

interest (that you need to control concentration, as in this case thyroid hormones). Also, FBS contains many different factors that may include the metabolite(s) of interest, such as thyroid hormones. For our studies, thyroid hormones depleted-FBS would be required to control thyroid hormones concentration. Depleted FBS can be obtained by activated charcoal and/or anion exchange resin treatment [15]. This procedure, however, also reduces other hormones and vitamins [16].

2. If mutant animals that require genotyping are being used, collect the waste tissues at this step (CNS, gastric tissue) for DNA extraction.
3. It is of utmost importance to use low passage MEFs (passage 2 to 4), as each passage will significantly decrease the number of reprogramming cells and delay the appearance of the first iPSC colonies. If frozen MEFs are to be used, freeze MEFs at passage 1 or 2, as freeze-thaw cycle will induce an important reduction of the reprogramming efficiency.
4. Of the canonical four Yamanaka factors, c-Myc can be omitted and still achieve iPSC reprogramming, although at a lower rate and more slowly.
5. We recommend to use fresh retroviral supernatants to perform the infection, as freezing the supernatants could reduce the viral titer. If freezing the retroviral suspension is needed, store at −80 °C and check viral titer to ensure proper infectivity rate.
6. The initial culture density should have to be determined based on the rate of cell proliferation (due to the cell line status or to variability of culture conditions). The cells should stand during the retroviral production without reaching confluence.
7. We recommend using FuGene 6 due to its ability of high performance in the presence of serum. Other reagents could also be used for transfecting the packaging cells, by optimizing the protocol.
8. Minimize the volume of DNA (that should be resuspended in water) to be used. Preferably, use a plasmid concentration around 1 μg/μL or higher.
9. To stablish an adequate number of infection rounds, a balance between infection efficiency and maintaining healthy MEFs should be achieved. This would depend on the viral titer and MEF conditions and resistance to infection. In our hands, using wild-type C57BL/6 derived MEFs, three rounds are adequate when the viral production efficiency is high.
10. In our hands, changing the medium every other day also works for the first 5 to 7 days of reprogramming. Afterward, the LIF is consumed fast requiring daily medium changes. This factor is critical for reprogramming and maintaining pluripotency.

11. For treatment with T3, due to T3 instability in solution, use a fresh aliquot of T3 each day to add to the iPSC Medium.
12. Automated counting could also be performed after plate scanning and applying the appropriate software solution.
13. This step is required if FBS is used. When using serum replacement this step is not necessary, as trypsin will not be inactivated, but could help to view the colonies.
14. Pick up as many colonies as possible within 15 min (around 6–10 per condition). Otherwise, the trypsin will evaporate and become concentrated. When planning to pick a high number of colonies, trypsin could be diluted 1:3 to prevent cell damage.
15. High-quality RNA is required for qRT-PCR, and that means A260/A280 ratio around 2.0, 260/230 ratio around 2.2, and RIN $\geq$ 8.

References

1. Takahashi K, Yamanaka S (2006) Induction of pluripotent stem cells from mouse embryonic and adult fibroblast cultures by defined factors. Cell 126(4):663–676. https://doi.org/10.1016/j.cell.2006.07.024
2. Takahashi K, Tanabe K, Ohnuki M et al (2007) Induction of pluripotent stem cells from adult human fibroblasts by defined factors. Cell 131(5):861–872. https://doi.org/10.1016/j.cell.2007.11.019
3. Park J, Kwon YW, Ham S et al (2017) Identification of the early and late responder genes during the generation of induced pluripotent stem cells from mouse fibroblasts. PLoS One 12(2):e0171300. https://doi.org/10.1371/journal.pone.0171300
4. Krause MN, Sancho-Martinez I, Izpisua Belmonte JC (2016) Understanding the molecular mechanisms of reprogramming. Biochem Biophys Res Commun 473(3):693–697. https://doi.org/10.1016/j.bbrc.2015.11.120
5. Boward B, Wu T, Dalton S (2016) Concise review: control of cell fate through cell cycle and pluripotency networks. Stem Cells 34(6):1427–1436. https://doi.org/10.1002/stem.2345
6. Polo JM, Hochedlinger K (2010) When fibroblasts MET iPSCs. Cell Stem Cell 7(1):5–6. https://doi.org/10.1016/j.stem.2010.05.018
7. Gonzales KA, Ng HH (2013) Driving pluripotency and reprogramming: nuclear receptors at the helm. Semin Cell Dev Biol 24(10–12):670–678. https://doi.org/10.1016/j.semcdb.2013.07.002
8. Zhuang Q, Li W, Benda C et al (2018) NCoR/SMRT co-repressors cooperate with c-MYC to create an epigenetic barrier to somatic cell reprogramming. Nat Cell Biol 20(4):400–412. https://doi.org/10.1038/s41556-018-0047-x
9. Ebrahimi A, Sevinç K, Gürhan Sevinç G et al (2019) Bromodomain inhibition of the coactivators CBP/EP300 facilitate cellular reprogramming. Nat Chem Biol 15(5):519–528. https://doi.org/10.1038/s41589-019-0264-z
10. Cheng SY, Leonard JL, Davis PJ (2010) Molecular aspects of thyroid hormone actions. Endocr Rev 31(2):139–170. https://doi.org/10.1210/er.2009-0007
11. Astapova I (2016) Role of co-regulators in metabolic and transcriptional actions of thyroid hormone. J Mol Endocrinol 56(3):73–97. https://doi.org/10.1530/JME-15-0246
12. Contreras-Jurado C, Montero-Pedrazuela A, Pérez RF et al (2023) The thyroid hormone enhances mouse embryonic fibroblasts reprogramming to pluripotent stem cells: role of the nuclear receptor corepressor 1. Front Endocrinol (Lausanne) 14:1235614. https://doi.org/10.3389/fendo.2023.1235614
13. Conner DA (2001) Mouse embryo fibroblast (MEF) feeder cell preparation. Curr Protoc Mol Biol. Chapter 23: Unit 23.2. https://doi.org/10.1002/0471142727.mb2302s51

14. Rao X, Huang X, Zhou Z, Lin X (2013) An improvement of the 2^(-delta delta CT) method for quantitative real-time polymerase chain reaction data analysis. Biostat Bioinforma Biomath 3:71–85
15. Samuels HH, Stanley F, Casanova J (1979) Depletion of L-3,5,3′-triiodothyronine and L-thyroxine in euthyroid calf serum for use in cell culture studies of the action of thyroid hormone. Endocrinology 105(1):80–85. https://doi.org/10.1210/endo-105-1-80
16. Cao Z, West C, Norton-Wenzel CS et al (2009) Effects of resin or charcoal treatment on fetal bovine serum and bovine calf serum. Endocr Res 34(4):101–108. https://doi.org/10.3109/07435800903204082

Chapter 9

DNA Methylation Analysis by Bisulfite Pyrosequencing of Mouse Embryonic Fibroblasts with Reprogramming Enhanced by Thyroid Hormones

Pablo Santamarina-Ojeda, Agustín F. Fernández, Mario F. Fraga, and Raúl F. Pérez

Abstract

DNA methylation is a widely studied epigenetic mark which in mammals involves the incorporation of a methyl group to the fifth carbon of cytosines, mainly those belonging to CpG dinucleotides. It has been linked to context-dependent regulatory functions ranging from gene and repetitive DNA silencing to gene body transcriptional activity. Because of its important roles during embryonic development and cell differentiation, DNA methylation can be used to track cell reprogramming by measuring the methylation levels of pluripotency-associated factors. In this scenario, bisulfite pyrosequencing is a simple, robust, and widely used technique which allows for the quantification of DNA methylation levels at small, specific regions of the genome. It involves the amplification and biotin tagging of bisulfite-converted DNA. Single amplified strands are then purified using streptavidin and finally pyrosequenced using a sequencing primer. Thus, it is an ideal method for the quantitative profiling of specific genomic regions, with applications ranging from biomarker discovery and epigenetic clock tracking to omic validation studies.

Key words DNA methylation, Epigenetics, Reprogramming, Pyrosequencing, Bisulfite, Thyroid hormones

1 Introduction

DNA methylation (DNAm) is an epigenetic mechanism which, in mammals, consists in the addition of a methyl group to the fifth carbon of cytosines, typically occurring within CpG dinucleotides. The genomic distribution and function of 5-methylcytosine (5mC) is context-dependent: CpG sites are either scattered throughout the genome or concentrated in "CpG islands", which are typically unmethylated and whose methylation is sometimes associated with gene repression [1], for example of X-chromosome inactivated genes or imprinted elements [2]. DNAm is also linked to the repression and stabilization of other genomic elements such as

Constanza Contreras-Jurado (ed.), *Thyroid Hormones: Methods and Protocols*, Methods in Molecular Biology, vol. 2876,
https://doi.org/10.1007/978-1-0716-4252-8_9,

heterochromatin and repetitive DNA [3]. On the other hand, this epigenetic mark can also correlate with gene activation, as is the case of gene-body methylation levels [4].

It is well-known that DNAm plays important roles in all core biological processes, ranging from development and cell differentiation [5] to aging and cancer [6, 7]. Indeed, 5mC is tightly regulated during embryonic development, when there occur "waves" of epigenetic reprogramming crucial for physiological development [2], while the mutation of DNA methyltransferases (DNMTs) leads to embryonic or postnatal lethality [8, 9]. Thus, it is no surprise that the study of DNAm can bring great insights to the field of cellular reprogramming [10–12]. For instance, it can be used to monitor reprogramming status by profiling the DNAm levels of pluripotency-associated factors such as *Oct4*, *Nanog*, or *Rex1* [13, 14].

Over the last five decades, a great number of methods have been developed for the detection, quantification, and genome-wide mapping of 5mC, ranging from chromatography assays and PCR-based methods to omic-scale microarrays and next-generation sequencing technologies [15], as well as state-of-the-art nanopore-sequencing methods [16–18]. In 1992, the use of sodium bisulfite as an effective strategy for the sequencing discrimination of methylated from unmethylated cytosines was first described [19, 20] and proved to be a major breakthrough which is now utilized by most of the gold standard methods in the field [21]. In essence, this reagent can deaminate unmethylated cytosine into uracil while leaving 5mC unaffected; then, by amplifying bisulfite-converted DNA using PCR, uracils are replaced by thymines, so that they can be finally distinguished from cytosines (originally methylated) by different methods [15].

In this setting, bisulfite pyrosequencing is a simple "sequencing by synthesis" technique which allows for the quantification of local DNAm levels at small regions of the genome (typically around 60–200 bp). The process involves the design of a triad of primers: two for the amplification plus biotin tagging of the target region and one for subsequent sequencing of the biotinylated strands. The protocol can be broken down into various parts (Fig. 1): (1) DNA is extracted and bisulfite-modified to transform unmethylated C into U; (2) the target region is amplified with biotin-tagged primers, so that amplified and biotinylated DNA is generated in which unmethylated C has been replaced with A; (3) the biotin-tagged strands are isolated using streptavidin beads and hybridized to the sequencing primer; (4) the strands are pyrosequenced by the sequential incorporation of nucleotides using a polymerase and detection of released pyrophosphate (PPi) via enzymatic reactions involving the generation of light [22, 23].

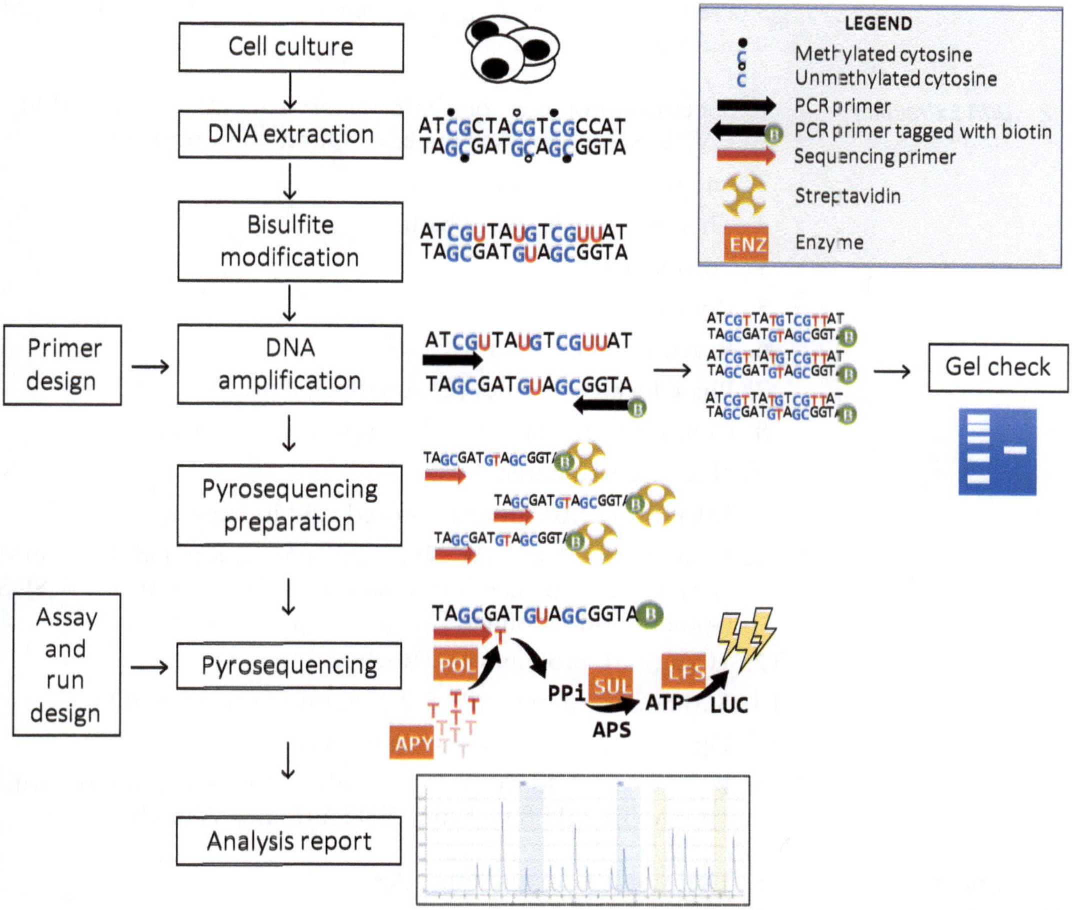

Fig. 1 Diagram indicating the different steps of the bisulfite pyrosequencing protocol. APY, apyrase; POL, DNA polymerase; SUL, ATP sulfurylase; LFS, luciferase; PPi, pyrophosphate; APS, adenosine phosphosulfate; ATP, adenosine triphosphate; LUC, luciferin

Today, bisulfite pyrosequencing is widely used in many applications, such as the discovery of biomarkers for cancer classification or treatment response [24, 25], the development of epigenetic clocks [26], the validation of results from genome-wide technologies such as DNAm arrays or whole-genome bisulfite sequencing (WGBS) [27], and the aforementioned uses in the monitoring of somatic cell reprogramming [13, 14].

2 Materials

2.1 Primer and Assay Design

1. Design of amplification plus sequencing primers: Pyromark™ Assay Design software (Qiagen).
2. Design of pyrosequencing assay: Pyromark™ Q24 software (Qiagen).

3. Preparation of each pyrosequencing run: Pyromark™ Q24 software (Qiagen).

2.2 DNA Extraction

1. Spectrophotometer or fluorometer: e.g., Nanodrop 1000™ UV/VIS Spectrophotometer or Qubit™ fluorometer.
2. Lab refrigerator and freezer.
3. Thermomixer or water bath.
4. Vortex mixer.
5. Rotary mixer.
6. Reaction tubes, 1.5 mL.
7. Phase lock gel™ tubes (QuantaBio), 2 mL.
8. Centrifuge, preferably with temperature control.
9. Nuclease-free water.
10. Ethanol at 100% and 75% stored cold in freezer.
11. Lysis buffer: 100 mM NaCl, 20 mM Tris pH 8.0, 25 mM EDTA (ethylenediaminetetraacetic acid) pH 8.0, 0.5% SDS (sodium dodecyl sulfate).
12. 10 mg/mL proteinase K (Roche).
13. 3 M Sodium acetate pH 5.2 (Thermo Fisher Scientific).
14. Optional: 20 mg/mL glycogen (Avantor).
15. Phenol:Chloroform:Isoamyl Alcohol (25:24:1) saturated with 10 mM Tris pH 8.0, 1 mM EDTA (Sigma-Aldrich).

2.3 Bisulfite Conversion

1. Thermomixer or thermocycler.
2. Centrifuge.
3. PCR reaction tubes, 200 μL.
4. Bisulfite conversion kit: Epitect Bisulfite Kit (Qiagen) or kits from other sources (e.g., EZ DNA Methylation-Gold kit from Zymo Research).
5. Methylated and unmethylated DNA standards (e.g., Methylation Controls from EpigenDx).

2.4 Alternative Conversion: Oxidative Bisulfite Conversion

1. Thermomixer or thermocycler.
2. Centrifuge.
3. DiaMag magnetic rack (Diagenode).
4. PCR reaction tubes, 200 μL.
5. Reaction tubes, 1.5 mL.
6. Nuclease-free water.
7. Agencourt AMPure XP magnetic beads (Beckman Coulter).
8. Acetonitrile 80% (Merck Millipore).

9. 1 M NaOH.
10. 375 mM $KRuO_4$ (Alfa Aesar) in 300 mM NaOH (*see* **Note 1**).
11. Bisulfite conversion kit: Epitect Bisulfite Kit (Qiagen).
12. Methylated and unmethylated DNA standards (e.g., Methylation Controls from EpigenDx).

2.5 DNA Amplification

1. Thermocycler.
2. PCR reaction tubes, 200 μL.
3. Nuclease-free water.
4. 10 μM PCR amplification primers, one of them biotinylated (*see* **Note 2**).
5. DNA polymerase and buffer: IMMOLASE™ DNA polymerase, 50 mM $MgCl_2$, 10x ImmoBuffer (Bioline).
6. 2 mM dNTP mix (Merck Millipore).

2.6 DNA Amplification Gel Check

1. Electrophoresis chamber.
2. Power supply.
3. Transilluminator.
4. Gel casting tray and comb.
5. Microwave or heating plate.
6. 2% Agarose in 0.5x TBE buffer (45 mM Tris, 45 mM Boric acid, 1 mM EDTA pH 8.3).
7. SimplySafe™ nucleic acid stain (EURx).
8. 5x loading buffer: 1.8 mM bromophenol blue, 273.5 mM sucrose.
9. GeneRuler 50 bp DNA ladder (Thermo Fisher Scientific).

2.7 Pyrosequencing

1. Orbital shaker.
2. Heating block.
3. 96-well flexible PCR plate.
4. 96-well plate cover foil adhesive.
5. PyroMark™ Q24 MDx pyrosequencer (Qiagen).
6. PyroMark™ Q24 Vacuum Prep Workstation with filter prep tool connected to a vacuum pump (Qiagen).
7. PyroMark™ Q24 Plates (Qiagen).
8. PyroMark™ Q24 Plate Holder (Qiagen).
9. PyroMark™ Q24 Cartridge (Qiagen).
10. Purified water.
11. 70% ethanol.
12. 3 μM sequencing primer.

13. Streptavidin Sepharose™ High Performance beads (Cytiva).
14. Pyromark™ Binding Buffer (Qiagen).
15. PyroMark™ Annealing Buffer (Qiagen).
16. PyroMark™ Denaturation Solution (Qiagen).
17. PyroMark™ Wash Buffer (Qiagen).
18. PyroMark™ Gold Q24 Reagents (Qiagen), containing enzymes mixture, substrate mixture, and dNTPs.

3 Methods

The bisulfite pyrosequencing protocol consists of a series of steps (*see* Fig. 1):

1. Design of pyrosequencing primers and pyrosequencing assay.
2. DNA extraction.
3. Bisulfite conversion.
4. PCR amplification and agarose gel check.
5. Pyrosequencing.

Note that the protocol here described is based on the Pyromark™ Q24 platform, and some differences will be encountered, particularly in the last steps, when using other platforms such as Pyromark™ Q24 Advanced, Pyromark™ Q48 Autoprep, or Pyromark™ Q96.

3.1 Primer and Assay Design

The primer and assay design involves the following steps:

1. Design of the three primers used in bisulfite pyrosequencing: two amplification primers, one tagged with biotin, and a sequencing primer.
2. Design of the pyrosequencing assay: this consists of creating a software file detailing the region to be analyzed, pyrosequencing reaction parameters, etc.

3.1.1 Primer Design

We typically use the Pyromark™ Assay Design software to design the primers because of its ease of use and its ability to design all three primers. Briefly, the user inputs the genomic sequence of interest, sets the "Assay type" to "Methylation analysis (CpG)", selects a target region, and proceeds with "Run Assay Design". Then, a list of the top sets of primers, sorted by score, will be shown to the user so that the selected set can be exported. Take note of the following information and recommendations:

1. When obtaining the target region genomic sequence, be aware that pyrosequencing is strand-specific, so the forward or reverse strand may be chosen.

2. The score of the software is quite stringent and sets of primers with scores as low as 60 may work properly.
3. The software provides a graphic view to inspect the primer positions and takes note of whether the biotin-tagged primer is the forward or the reverse primer as this is important during the assay design step.
4. If no good-quality primers are found, the user may allow the design of "primers over variable positions", i.e., primers containing CpG sites. These primers will degenerate and contain a mixture of oligonucleotides covering the possible sequences resulting from the bisulfite conversion of the methylated or unmethylated CpG sites (*see* **Note 2**).
5. Degenerate forward primers should be ordered with a Y base on the variable position, while reverse primers should have an R base.

3.1.2 Assay Design

We use the Pyromark™ Q24 2.0.6 software to design the pyrosequencing assay and export subsequent runs. Briefly, the user selects a "New assay: CpG assay" and introduces the bisulfite-converted sequence into the software. Then, the software will generate a nucleotide dispensation sequence based on the input genomic sequence, which will be read by the pyrosequencer to perform the sequencing reaction. Take note of the following information and recommendations:

1. To prepare the bisulfite-converted sequence: retain only the sequence starting from the end of the sequencing primer and successively substitute all CG for YG and then all C for T.
2. If the biotin-tagged amplification primer is the forward primer, the sequencing primer will be a reverse primer, and the region input to the software should be the reverse complement. This also impacts the bisulfite controls selected in the software.
3. The quality of the pyrosequencing signal decreases during the pyrosequencing run so that a significant quality drop is often observed after 60–80 bp. Thus, optimal results will be obtained for sequences below those lengths.
4. The software will show a graphical histogram detailing the nucleotide sequence and dispensation order: note that each CpG site includes a variable position (C or T) so both nucleotides will be sequentially dispensed to evaluate DNA methylation levels.
5. An important point is the incorporation of "bisulfite controls": these are points in the dispensation order in which cytosine is dispensed at locations in which there are thymines which originally came from non-CpG cytosines prior to bisulfite

treatment, so that ideally, an absence of cytosine signal is to be expected at those locations.

3.2 Pyrosequencing Run Export

After creating an assay, any number of pyrosequencing runs can be exported for a specific assay. Exporting a pyrosequencing run will return a "pyrorun" file to input into the pyrosequencer and also a "pre-run" document listing the necessary quantities of PyroMark™ Gold Q24 Reagents (nucleotides, enzyme, and substrate mixtures) to be used during the pyrosequencing.

Take note of the following information and recommendations:

1. The appropriate instrument method identifier (typically indicated in the PyroMark™ Q24 Cartridge) should be specified.
2. The plate set-up will show up as a 24-well grid: different assays may be selected to be pyrosequenced in different wells.

3.3 Pyrosequencing Assay Controls

If creating a pyrosequencing assay for the first time, a series of controls must be specified in some of the plate wells of the first pyrosequencing run to ensure proper functionality:

1. Amplified PCR product, PyroMark™ Annealing Buffer and no sequencing primer.
2. Water PCR control, PyroMark™ Annealing Buffer and sequencing primer.
3. PyroMark™ Annealing Buffer and sequencing primer, without any PCR product (2.5 μL of 3 μM sequencing primer, 22.5 μL of annealing buffer).
4. PyroMark™ Annealing Buffer and biotinylated primer, without any PCR product (0.75 μL of 10 μM biotinylated primer, 24.25 μL of annealing buffer).
5. PyroMark™ Annealing Buffer, biotinylated primer and sequencing primer, without any PCR product (2.5 μL of 3 μM sequencing primer, 0.75 μL of 10 μM biotinylated primer, 21.75 μL of annealing buffer).

3.4 DNA Extraction

For bisulfite pyrosequencing we typically employ a phenol-based extraction protocol because it is versatile and gives a substantial DNA yield with adequate quality for subsequent processing [28]:

1. Add 500 μL of lysis buffer to the sample and homogenize.
2. Add 50 μL of 10 mg/mL proteinase K and vortex.
3. Incubate at 55 °C overnight in the thermomixer with medium shaking (e.g., 500 rpm). For cell extracts, 1–2 h incubation may be enough.
4. Inactivate proteinase K by incubating at >70 °C for 15 min.

5. Centrifuge the phase lock gel tube (>10,000 *g* × 20 s) to pellet the gel and transfer sample to the tube.
6. Add 500 μL of Phenol:Chloroform:Isoamyl Alcohol.
7. Mix for 30 min at room temperature in the rotary mixer.
8. Centrifuge at >10,000 *g* × 5 min.
9. Transfer the upper phase (aqueous) to a new 1.5 mL reaction tube.
10. Add 50 μL of room temperature 3 M sodium acetate pH 5.2 and mix by inverting the tube. Optionally, 1 μL of 20 mg/mL glycogen may be added to aid DNA precipitation.
11. Add 2 volumes of cold 100% ethanol (around 1000 μL) and mix by inverting the tube.
12. Incubate for at least 4–6 h at −20 °C (*see* **Note 3**).
13. Centrifuge at >10,000 *g* × 20 min (and preferably at 4 °C), to pellet the DNA.
14. Discard the supernatant and add 1 mL of cold 75% ethanol to wash the pellet.
15. Centrifuge at >10,000 *g* × 5 min (and preferably at 4 °C) to pellet the clean DNA again.
16. Optional: repeat **steps 14** and **15.**
17. Carefully discard the supernatant and air-dry the pellet at room temperature.
18. Resuspend the pellet in 25–100 μL of nuclease-free water or other buffers. Incubating the sample for 15 min at 37 °C may help with the resuspension.
19. Quantify the extracted DNA with a spectrophotometer or fluorometer.

3.5 Bisulfite Conversion

In this critical step, DNA is modified using sodium bisulfite to transform cytosines into uracils while leaving 5mC unaffected. Follow the manufacturer's instructions for the Epitect Bisulfite Kit (Qiagen) and take note of the following recommendations:

1. We recommend starting from at least 200 ng of DNA (optimally 500 ng).
2. We recommend handling the samples using 200 μL PCR reaction tubes.
3. We strongly recommend the use of methylated and unmethylated DNA standards, which are treated as independent samples and must be bisulfite-converted (*see* **Note 4**).
4. We typically elute in 20 μL of nuclease-free water and use 2 μL in the subsequent PCR amplification step.

5. Bisulfite-converted DNA is less stable than non-treated DNA but can be stored for 1 to 5 days at 4 °C and for several weeks preferably at −80 °C. Repeated freeze-thawing should be avoided.

3.6 Alternative Conversion: Oxidative Bisulfite Conversion

5-Hydroxymethylcytosine (5hmC) is an epigenetic mark which arises from the oxidation of 5mC by action of the ten-eleven translocation (TET) enzymes. This mark exists in very low levels across most tissues [29, 30] but is known to correlate with genomic functions [31] so it may be desired to measure its levels. Bisulfite cannot distinguish 5hmC from 5mC, so bisulfite-based quantifications of 5mC levels actually measure the sum of these two epigenetic marks. To be able to distinguish them, an oxidation step using potassium perruthenate ($KRuO_4$) can be performed prior to bisulfite conversion which allows for the direct quantification of 5mC [32]. Subtracting these "true" 5mC values from "standard" bisulfite-converted values (5mC + 5hmC) will give the levels of 5hmC (Fig. 2, *see* **Note 5**). This oxidation is carried out before bisulfite conversion and consists of the following steps:

1. Start from a minimum of 2000 ng of isolated DNA adjusted to a volume of 50 μL in a 1.5 mL reaction tube.
2. Add 90 μL of magnetic beads, mix thoroughly, and incubate at room temperature for 20 min.
3. Centrifuge at >10,000 g x 1 min to pellet the bead-DNA complexes.
4. Place the tube in the magnetic rack so that the beads bind to the inner wall of the tube.
5. Remove the supernatant and wash three times with acetonitrile 80%. It's critical to not pour the liquid directly on the beads and to avoid directly contacting the beads with the pipette tip.

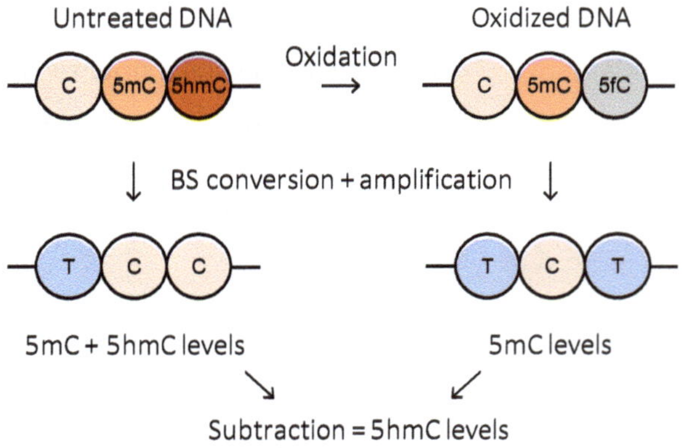

Fig. 2 Diagram showing the process of bisulfite conversion or oxidative bisulfite conversion of DNA. 5mC, 5-Methylcytosine; 5fC, 5-Formylcytosine; 5hmC, 5-Hydroxymethylcytosine

6. Air-dry the open tube for 10 min.
7. Add 50 μL of 1 M NaOH, mix thoroughly, and incubate at 37 ° C for 30 min.
8. Centrifuge at >10,000 g x 1 min to pellet the beads.
9. Place the tube in the magnetic rack so that the beads bind to the inner wall of the tube.
10. Divide the volume at the bottom of the 1.5 mL tubes (around 25 μL per half) into two distinct 200 μL PCR reaction tubes: labeled as "BS" and "oxBS" depending on whether the oxidation will be performed or not.
11. Add 1 μL of 375 mM $KRuO_4$ in 300 mM NaOH to the "oxBS"-labeled tube and an equivalent volume of water to the "BS"-labeled tube.
12. Incubate at 40 °C for 30 min (*see* **Note 6**).
13. Centrifuge at >10,000 g × 10 min and recover the supernatant.
14. Proceed with bisulfite conversion following the indications of the Epitect Bisulfite Kit (*see* **Note 7**).
15. We typically elute in 20 μL of nuclease-free water and use 2 μL in the subsequent PCR amplification step.

3.7 DNA Amplification

In this part of the protocol, we amplify the bisulfite-converted DNA so that uracils (originally unmethylated cytosines) are replaced by thymines in the final amplified product. In this step, one of the PCR amplification primers used is tagged with biotin so that one of the amplified strands can be subsequently captured using streptavidin beads to be subsequently pyrosequenced.

1. Prepare the following 28 μL PCR mix:

Nuclease-free water	20.3 μL
ImmoBuffer 10x	3 μL
dNTP mix (2 mM)	3 μL
$MgCl_2$ (50 mM)	0.9 μL
Primer F (10 μM)	0.3 μL
Primer R (10 μM)	0.3 μL
IMMOLASE™ DNA polymerase	0.2 μL

2. We typically add 2 μL of the bisulfite-converted DNA which we obtain when treating an initial quantity of 200–500 ng and eluting in 20 μL in the previous step.
3. Run the following program in the thermocycler (*see* **Note 8**):

95 °C	10 min	Initial denaturation
95 °C Melting temperature for the primers 72 °C	30 s 30 s 30 s	50 cycles
72 °C	7 min	Final extension

4. The amplified product can be stored for 1 to 5 days at 4 °C or for several weeks at -20 °C.

3.8 DNA Amplification Gel Check

In this step we check if the bisulfite-converted DNA was successfully amplified during the PCR, while also ensuring that negative controls do not show any amplification. Take note of the following recommendations:

1. We typically use 5 μL of the 30 μL PCR product for the electrophoresis check and mix with 1–1.2 μL of 5x loading buffer.
2. We prepare 2% agarose gels in 0.5x TBE and add 2 μL of SimplySafe™ nucleic acid stain per 50 μL of gel.
3. We typically run the gel at 90–100 V for 15 min.

3.9 Pyrosequencing

This is the final step of the protocol: on the one hand, biotinylated single DNA strands are captured from the previously amplified product using streptavidin-coated beads and are subsequently annealed to the pyrosequencing primer; on the other hand, pyrosequencing reagents (enzymes, substrate, dNTPs) are prepared in a cartridge and introduced into the machine. Finally, the purified strands are sequenced following a pyrosequencing run designed on the Pyromark™ Q24 software.

1. Prepare the pyrosequencing run on the Pyromark™ Q24 software and take note of the required quantities of dNTPs, enzyme and substrate mixture.
2. Turn on the PyroMark™ Q24 pyrosequencer, introduce the pyrosequencing run and leave on standby.
3. Let the PyroMark™ Q24 Gold reagents reach room temperature, carefully fill the PyroMark™ Q24 Cartridge avoiding air bubbles, and place it in the pyrosequencer (*see* **Note 9**).
4. Pre-heat a heating block to 80 °C and place the PyroMark™ Q24 Plate Holder on top.

5. Prepare a 24-well PCR plate and transfer 20 μL of the PCR product to each well (*see* **Note 10**). The sample arrangement must match that disposed in the pyrosequencing run.
6. Gently resuspend the Sepharose™ beads (no vortexing) and add 40 μL of Pyromark™ Binding Buffer, 19 μL of water and 1 μL of Sepharose™ beads to the PCR product to reach a final volume of 80 μL.
7. Cover the plate with plate cover adhesive and leave shaking at room temperature and 2000 rpm for at least 10 min. The orbital shaker revolutions must be enough to prevent sedimentation of the beads.
8. During the shaking step, prepare the PyroMark™ Q24 Plate with 22.5 μL of Pyromark™ Annealing Buffer and 2.5 μL of 3 μM sequencing primer. The reagent arrangement in the plate must resemble that disposed in the run and also in the shaking PCR plate.
9. Prepare the PyroMark™ Q24 Vacuum Prep Workstation by filling the troughs with 70%, ethanol, purified water, PyroMark™ Denaturation Solution and PyroMark™ Wash Buffer.
10. Turn the vacuum on and flush the filter probes from the workstation with purified water. Refill the depleted trough and leave the filter on standby. Make sure that none of the aspirating filters are clogged.
11. Take the PCR plate from the orbital shaker, immediately place it on the workstation and aspirate the bead mixture (15 s), carefully checking that all the volume has been aspirated in all wells. Avoid delays in this step to prevent bead sedimentation. You may be able to see the beads successfully retained on the tips of the filters.
12. Successively aspirate ethanol (5 s), PyroMark™ Denaturation Solution (5 s), and PyroMark™ Wash Buffer (10 s) (*see* **Note 11**).
13. Raise the filter tips at an angle exceeding 90 degrees and turn of the vacuum until it is completely released.
14. Place the filters on the PyroMark™ Q24 Plate to release the beads onto the wells. Shake mildly for 30–60 s.
15. Remove the filters, check that the volume of the PyroMark™ Q24 Plate has not been accidentally aspirated (*see* **Note 12**), and place the filters in the waste water trough.
16. Shake the filters in the waste water trough without aspiration to remove excess beads, turn on the vacuum, and fully aspirate the next clean water trough to clean the filters.
17. Incubate the PyroMark™ Q24 Plate at 80 °C for 2 min on the pre-heated PyroMark™ Q24 Plate Holder.

18. Let the PyroMark™ Q24 Plate reach room temperature (3–5 min) and place it in the pyrosequencer.
19. Start the pyrosequencing run and analyze the results on the Pyromark™ Q24 software.
20. Consecutive pyrosequencing runs may be prepared by adding PyroMark™ Q24 Gold reagents to the same cartridge.
21. During the last pyrosequencing run, clean the PyroMark™ Q24 Vacuum Prep Workstation. Plastic trays may be rinsed with purified water and reutilized.
22. After finishing the last pyrosequencing run, clean the cartridge by gently removing the excess dNTPs and enzyme and substrate mixtures. Carefully rinse all the cartridge wells with water and, for the enzyme and substrate mixture tips, apply pressure with your fingertips until water is observed passing through the tips. Store the cartridge upside down taking care not to expose the delicate tips.

4 Notes

1. We usually keep a stock of $KRuO_4$ at −20 °C and prepare the fresh solution of 375 mM $KRuO_4$ in 300 mM NaOH when needed.
2. Biotinylated primers should be ordered with HPLC purification and always stored at −20 °C. Additionally, note that if degenerate primers are employed these might have to be prepared at higher concentrations to achieve the same molarity as the standard primers.
3. An incubation of 2–3 h at −80 °C can also be enough to precipitate DNA.
4. Methylated and unmethylated DNA standards can be used to detect PCR bias due to unequal amplification of bisulfite-converted DNA. This biased amplification is caused by the generation of molecules with sequence differences after bisulfite-conversion due to their differing initial methylation statuses [33].
5. If one is not interested in quantifying 5hmC, we do not suggest using the oxidation method to measure "true" 5mC levels because it is much more damaging to the DNA samples and it is not standard practice in the field.
6. At this point in the protocol, the oxidation reaction should manifest with a black-orange color. This can help calibrate the optimal amount of $KRuO_4$ used.
7. After conversion, it can be assumed that oxidized-plus bisulfite-converted DNA is even less stable than bisulfite-converted

DNA. We recommend short-term storage, preferably at −80 °C, avoiding repeated freeze-thawing.

8. We typically use 50 cycles to ensure strong amplification of the product for pyrosequencing, but this number can often be reduced to avoid unwanted amplifications. Always include several negative controls in the PCR reaction to check for undesired amplifications and contamination.
9. Micropipette tips with no filters are highly recommended to avoid clogging the cartridge.
10. It is useful to mark a corner of the 24-well plate to keep track of its orientation.
11. Always make sure that aspirated fluid is observed passing through the vacuum hose prior to counting the aspiration times. This may take some seconds if the filters are partially clogged.
12. If the vacuum is not properly turned off and the PyroMark™ Q24 Plate contents are accidentally aspirated, a new plate with Pyromark™ Annealing Buffer and sequencing primer may be prepared in which to release the beads still retained in the filter.

Acknowledgments

Funding This work received no external funding.

Conflict of Interest The authors declare no conflict of interest.

Author Contributions R.F.P. conceptualized and wrote the manuscript. P.S.O., A.F.F., and M.F.F. assisted in manuscript writing. All authors revised, read, and approved the final manuscript.

References

1. Edwards JR, Yarychkivska O, Boulard M, Bestor TH (2017) DNA methylation and DNA methyltransferases. Epigenetics Chromatin 10:23. https://doi.org/10.1186/s13072-017-0130-8
2. Greenberg MVC, Bourc'his D (2019) The diverse roles of DNA methylation in mammalian development and disease. Nat Rev Mol Cell Biol 20:590–607. https://doi.org/10.1038/s41580-019-0159-6
3. Pappalardo XG, Barra V (2021) Losing DNA methylation at repetitive elements and breaking bad. Epigenetics Chromatin 14:25. https://doi.org/10.1186/s13072-021-00400-z
4. Jones PA (2012) Functions of DNA methylation: islands, start sites, gene bodies and beyond. Nat Rev Genet 13:484–492. https://doi.org/10.1038/nrg3230
5. Atlasi Y, Stunnenberg HG (2017) The interplay of epigenetic marks during stem cell differentiation and development. Nat Rev Genet 18:643–658. https://doi.org/10.1038/nrg.2017.57

6. Feinberg AP, Koldobskiy MA, Göndör A (2016) Epigenetic modulators, modifiers and mediators in cancer aetiology and progression. Nat Rev Genet 17:284–299. https://doi.org/10.1038/nrg.2016.13
7. Sen P, Shah PP, Nativio R, Berger SL (2016) Epigenetic mechanisms of longevity and aging. Cell 166:822–839. https://doi.org/10.1016/j.cell.2016.07.050
8. Li E, Bestor TH, Jaenisch R (1992) Targeted mutation of the DNA methyltransferase gene results in embryonic lethality. Cell 69:915–926. https://doi.org/10.1016/0092-8674(92)90611-f
9. Okano M, Bell DW, Haber DA, Li E (1999) DNA methyltransferases Dnmt3a and Dnmt3b are essential for de novo methylation and mammalian development. Cell 99:247–257. https://doi.org/10.1016/s0092-8674(00)81656-6
10. Hu X, Zhang L, Mao S-Q et al (2014) Tet and TDG mediate DNA demethylation essential for mesenchymal-to-epithelial transition in somatic cell reprogramming. Cell Stem Cell 14:512–522. https://doi.org/10.1016/j.stem.2014.01.001
11. Lee D-S, Shin J-Y, Tonge PD et al (2014) An epigenomic roadmap to induced pluripotency reveals DNA methylation as a reprogramming modulator. Nat Commun 5:5619. https://doi.org/10.1038/ncomms6619
12. Greenberg MVC (2020) Get out and stay out: new insights into DNA methylation reprogramming in mammals. Front Cell Dev Biol 8:629068. https://doi.org/10.3389/fcell.2020.629068
13. Nishino K, Toyoda M, Yamazaki-Inoue M et al (2011) DNA methylation dynamics in human induced pluripotent stem cells over time. PLoS Genet 7:e1002085. https://doi.org/10.1371/journal.pgen.1002085
14. Contreras-Jurado C, Montero-Pedrazuela A, Pérez RF et al (2023) The thyroid hormone enhances mouse embryonic fibroblasts reprogramming to pluripotent stem cells: role of the nuclear receptor corepressor 1. Front Endocrinol (Lausanne) 14:1235614. https://doi.org/10.3389/fendo.2023.1235614
15. Harrison A, Parle-McDermott A (2011) DNA methylation: a timeline of methods and applications. Front Genet 2:74. https://doi.org/10.3389/fgene.2011.00074
16. Laszlo AH, Derrington IM, Brinkerhoff H et al (2013) Detection and mapping of 5-methylcytosine and 5-hydroxymethylcytosine with nanopore MspA. Proc Natl Acad Sci USA 110:18904–18909. https://doi.org/10.1073/pnas.1310240110
17. Rand AC, Jain M, Eizenga JM et al (2017) Mapping DNA methylation with high-throughput nanopore sequencing. Nat Methods 14:411–413. https://doi.org/10.1038/nmeth.4189
18. Simpson JT, Workman RE, Zuzarte PC et al (2017) Detecting DNA cytosine methylation using nanopore sequencing. Nat Methods 14:407–410. https://doi.org/10.1038/nmeth.4184
19. Frommer M, McDonald LE, Millar DS et al (1992) A genomic sequencing protocol that yields a positive display of 5-methylcytosine residues in individual DNA strands. Proc Natl Acad Sci USA 89:1827–1831. https://doi.org/10.1073/pnas.89.5.1827
20. Hayatsu H (2008) Discovery of bisulfite-mediated cytosine conversion to uracil, the key reaction for DNA methylation analysis–a personal account. Proc Jpn Acad Ser B Phys Biol Sci 84:321–330. https://doi.org/10.2183/pjab.84.321
21. Mattei AL, Bailly N, Meissner A (2022) DNA methylation: a historical perspective. Trends Genet 38:676–707. https://doi.org/10.1016/j.tig.2022.03.010
22. Tost J, Gut IG (2007) DNA methylation analysis by pyrosequencing. Nat Protoc 2:2265–2275. https://doi.org/10.1038/nprot.2007.314
23. Delaney C, Garg SK, Yung R (2015) Analysis of DNA methylation by pyrosequencing. Methods Mol Biol 1343:249–264. https://doi.org/10.1007/978-1-4939-2963-4_19
24. Chai R-C, Liu Y-Q, Zhang K-N et al (2019) A novel analytical model of MGMT methylation pyrosequencing offers improved predictive performance in patients with gliomas. Mod Pathol 32:4–15. https://doi.org/10.1038/s41379-018-0143-2
25. Rodríguez-Rodero S, Morales-Sánchez P, Tejedor JR et al (2022) Classification of follicular-patterned thyroid lesions using a minimal set of epigenetic biomarkers. Eur J Endocrinol 187:335–347. https://doi.org/10.1530/EJE-22-0012
26. Han Y, Eipel M, Franzen J et al (2018) Epigenetic age-predictor for mice based on three CpG sites. elife 7:e37462. https://doi.org/10.7554/eLife.37462
27. BLUEPRINT Consortium (2016) Quantitative comparison of DNA methylation assays for biomarker development and clinical applications. Nat Biotechnol 34:726–737. https://doi.org/10.1038/nbt.3605

28. Kopecká J, Matoulková D, Němec M et al (2014) Comparison of DNA extraction methods in terms of yield, purity, long-term storage, and downstream manipulation with Brewer's yeast chromosomal DNA. J Am Soc Brew Chem 72:1–5. https://doi.org/10.1094/ASBCJ-2014-0110-01
29. Szwagierczak A, Bultmann S, Schmidt CS et al (2010) Sensitive enzymatic quantification of 5-hydroxymethylcytosine in genomic DNA. Nucleic Acids Res 38:e181. https://doi.org/10.1093/nar/gkq684
30. Globisch D, Münzel M, Müller M et al (2010) Tissue distribution of 5-hydroxymethylcytosine and search for active demethylation intermediates. PLoS One 5:e15367. https://doi.org/10.1371/journal.pone.0015367
31. Cui X-L, Nie J, Ku J et al (2020) A human tissue map of 5-hydroxymethylcytosines exhibits tissue specificity through gene and enhancer modulation. Nat Commun 11:6161. https://doi.org/10.1038/s41467-020-20001-w
32. Booth MJ, Branco MR, Ficz G et al (2012) Quantitative sequencing of 5-methylcytosine and 5-hydroxymethylcytosine at single-base resolution. Science 336:934–937. https://doi.org/10.1126/science.1220671
33. Busato F, Dejeux E, El Abdalaoui H et al (2018) Quantitative DNA methylation analysis at single-nucleotide resolution by Pyrosequencing®. Methods Mol Biol 1708:427–445. https://doi.org/10.1007/978-1-4939-7481-8_22

Part III

Thyroid Hormones Receptors

Chapter 10

Analysis of the Pattern of RNA Expression in the Skin of TR-Deficient Mice By RNA-seq

María Gallardo-Gómez

Abstract

RNA-sequencing (RNA-seq) has become the method of choice for whole transcriptome analysis, as it enables profiling of a wide range of RNA molecules and the always evolving bioinformatic pipelines allow the extraction of diverse biological information from a single sample. However, there is not a gold-standard RNA-seq protocol nor an optimal bioinformatic pipeline defined, which can challenge new users. The aim of this chapter is to describe the basic RNA-sequencing pipeline, from RNA extraction and library preparation to the bioinformatic preprocessing and downstream analyses. The steps are oriented to the transcriptome analysis of the skin of TR-deficient mice, but the protocol can be easily translated to other tissue types or organisms.

Key words RNA-sequencing, Transcriptome profiling, Illumina, Bioinformatics, Differential expression analysis

1 Introduction

Although DNA microarrays are still in use, transcriptome profiling by next-generation sequencing (RNA-sequencing or RNA-seq) is currently the technology of choice for genome-wide gene expression quantification. Other than mRNA transcript quantification, RNA-seq is applicable to the profiling of total RNA, small RNA, long-non-coding RNA, and also to inferring strand-, isoform-, and variant-specific information [1, 2]. Also, the continuous development of bioinformatic algorithms allows the extraction of novel biological information from RNA-seq data, such as the identification of gene fusions, estimation of tissue cell type proportions, analysis of copy number alterations, or quantify the activation of transposable elements [3]. Such a broad scope is a unique feature of this technology compared to DNA microarrays or qPCR-based methods.

Constanza Contreras-Jurado (ed.), *Thyroid Hormones: Methods and Protocols*, Methods in Molecular Biology, vol. 2876, https://doi.org/10.1007/978-1-0716-4252-8_10,

A wide range of RNA-seq protocols have been published, there are several sequencing platforms available and the optimal bioinformatic pipeline is not defined [4], as data can be processed using different combinations of modular software. These factors make it challenging for naive users to appreciate an overview of all steps necessary to conduct or design an RNA-seq study. The basic workflow of RNA sequencing includes RNA purification, cDNA library preparation and sequencing, alignment of the sequencing data to the reference genome, and differential expression analysis.

The differential expression profiles identified by RNA-seq might be linked with changes in the homeostasis of tissues and organs giving an indication for further investigation. Previous work on TR knock-out mice has reported that several species of micro-RNAs such as miR-21, miR-31, miR34, and miR203 play a crucial role in the homeostasis of skin [5]. In this chapter we describe the basic procedure to analyze the whole skin transcriptome in TR-deficient mice by RNA-seq, to identify novel genes and molecular pathways that can be altered upon the depletion of thyroid receptors (TR). Here we cover the main steps of RNA-seq (Fig. 1), from tissue preservation to the obtention of curated gene expression patterns, which includes both laboratory work and bioinformatic analyses. The workflow presented in this chapter can be easily adapted to other tissue types or organisms.

2 Materials

2.1 Skin Tissue Preservation in RNAlater®

1. RNAlater® RNA stabilization solution (ThermoFisher).

2.2 RNA Extraction from Mice Skin Preserved in RNAlater® Solution

1. Total RNA extraction from animal tissue samples (for skin tissue we use RNeasy® Fibrous Tissue Mini kit from Qiagen), with ready-to-use thawed DNAse I stock solution in RNase-free water, β-mercaptoethanol or dithiothreitol added to buffer RLT Plus, 96–100% ethanol added to RPE buffer and all buffers equilibrated to room temperature (*see* **Note 1**).
2. Equipment for tissue disruption and homogenization.

2.3 RNA Quantification and Integrity Assessment

1. Fluorometer for RNA quantification (e.g., Qubit, ThermoFisher), with appropriate assay tubes and assay kit (e.g., Qubit RNA HS Assay or Qubit RNA BR Assay kit).
2. System for RNA quality and integrity evaluation with their corresponding kit, such as 4150 TapeStation System (RNA ScreenTape assay kit) or 2100 Bioanalyzer system (RNA 6000 Nano assay).

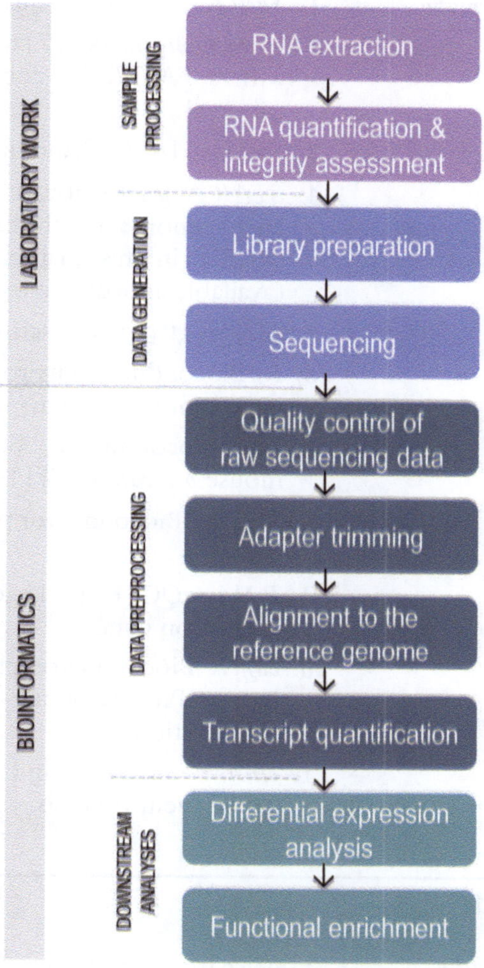

Fig. 1 Basic workflow of gene expression profiling with RNA-sequencing

2.4 Library Preparation and Library Size Assessment

1. Library preparation kit for mRNA with poly(A) tails such as TrueSeq RNA Sample Prep Kit v2 (Illumina).
2. System for cDNA library size and concentration assessment with their corresponding kit, such as 4150 TapeStation System (HS D1000 ScreenTape assay, Agilent) or 2100 Bioanalyzer system (HS DNA assay, Agilent).

2.5 Transcriptome Sequencing

1. A sequencing system that is compatible with your cDNA library (e.g., Illumina HiSeq2000 if a TrueSeq cDNA library is used).

2.6 Bioinformatic Analyses

1. *FastQC* tool [6] for quality control of raw sequencing data.
2. *Trimmomatic* tool [7] to remove sequencing adapters together with a FASTA file containing the adapter sequences to be trimmed.
3. *Samtools* [8] for handling SAM/BAM format files.
4. *STAR* Aligner software [9] with the most recent mouse reference genome (both FASTA and GTF files). The pipeline detailed in this chapter uses the mouse GRCm39 assembly (available at http://www.ensembl.org/).
5. *HTseq* [10] for transcript quantification.
6. R/Bioconductor environment [11] for downstream analysis of gene expression data.
7. The Bioconductor library of genome-wide annotation for mouse *org.Mm.eg.db* [12].
8. *DESeq2* Bioconductor package [13] for differential expression analysis.
9. *RNAseqQC* R package [14] for quality control of gene expression data.
10. *edgeR* Bioconductor package [15] to normalize count data into CPM (Counts per Million) for expression levels visualization.
11. *clusterProfiler* [16] and *enrichplot* [17] R packages for Gene Set Enrichment Analysis.

3 Methods

3.1 Tissue RNA Preservation in RNAlater®

1. Mice fresh skin sections no bigger than 0.5 cm are immersed in 5–10 volumes of RNAlater® solution. Do not freeze skin before immersion.
2. Store at 4 °C overnight to allow the solution to penetrate the tissue.
3. Remove the supernatant and move the sample to −20 °C. Mice skin tissue can be preserved in RNAlater® indefinitely at −20 °C (*see* **Note 2**).

3.2 RNA Extraction from Skin Tissue Preserved in RNAlater®

1. Sample preparation. Thaw preserved samples at room temperature. Remove the piece of skin from the RNAlater® using sterile forceps and be sure to remove any excess of reagents or any crystals/precipitates that may have formed during long-term storage.
2. Tissue disruption and homogenization. An efficient disruption and homogenization of the starting tissue are required for total

RNA purification. Thus, disrupt the skin tissue and homogenize the lysate following the instructions of your RNA extraction kit.

3. RNA extraction. Proceed with the total RNA extraction from the lysate following the steps detailed in your total RNA extraction kit. Briefly, if using RNeasy® Fibrous Tissue Mini kit, lysates are treated with Proteinase K and debris are pelleted by centrifugation. The supernatant is then cleared with 96–100% ethanol and centrifuged through a spin column, where RNA binds to the silica membrane. The silica membrane usually contains DNase that removes any possible traces of genomic DNA. After DNase and other contaminants are efficiently washed away, total RNA is eluted in RNase-free water (*see* **Note 3**).

3.3 RNA Quality Assessment and Quantification

1. Quantify RNA concentration. Proceed to RNA concentration quantification with the Qubit™ fluorimeter, following the manufacturer's instructions detailed in the Qubit RNA HS assay or Qubit RNA BR assay (*see* **Note 4**).
2. Check RNA integrity. The integrity of the RNA samples can be objectively evaluated by either a TapeStation or a Bioanalyzer, as both systems automatically provide quantitative data on RNA integrity based on the RNA integrity number (RIN). RIN values range from 10 (intact), 5 (partially degraded) to <3 (strongly degraded) [18, 19]. Unfortunately, RNA derived from FFPE tissues renders lower RIN values and a low RIN is not the best indicator of sample quality. Alternatively, it is possible to calculate the percentage of RNA fragments of more than 200 nt (DV_{200}) as a reliable indicator of FFPE RNA quality. Intact (non-FFPE) RNA usually has a $DV_{200} > 70\%$. A $DV_{200} > 50\%$ indicates good quality FFPE RNA, a DV_{200} ranging from 20 to 50% indicates poor FFPE RNA although it is suitable for further processing, while a sample with a $DV_{200} < 20\%$ is not recommended for further processing [20, 21].

3.4 Library Preparation and Library Size Assessment

1. Library preparation. Select a library that is compatible with the source of your RNA, as not all library preparation kits are compatible with FFPE tissue. Follow the manufacturer's instruction to construct your cDNA library: for the TruSeq RNA Library Prep Kit v2, mRNA (Poly(A) RNA) is enriched from total RNA, chemically fragmented, and converted into single-stranded complementary DNA (sscDNA) using random 6-mer primers, followed by the generation of double-stranded cDNA (dscDNA). Next, blunt-end fragments are created and sequencing adapters (which contain the full complement

sequence of sequencing primers) are ligated to the cDNA fragments which are amplified into cDNA libraries.

2. Library quality assessment and quantification. Quality and quantity of cDNA libraries must be assessed before RNA-sequencing in either a TapeStation or a Bioanalyzer system, following the protocol of the corresponding assay kit. In both platforms, a good quality cDNA library must report a peak of DNA fragment size between 150-300 bp in the electropherogram. The library concentration can be determined by integrating under the entire peak.

3.5 Transcriptome Sequencing

1. Paired-end sequencing. Libraries are applied to an Illumina flow cell for cluster generation within the corresponding library-compatible sequencing platform (e.g., Illumina HiSeq2000). Sequencing cycles are calculated based on the desired read length. For 100 bp paired-end (PE) sequencing, the Illumina sequencing instrument has to run for 200 cycles. Image analysis and per-cycle base calling is performed with the *Illumina Real Time Analysis* (RTA) software.
2. Raw data acquisition. Resulting sequencing data is converted into FASTQ format using the Illumina software *Bcl2Fastq*, which comes as standard in the Illumina sequencing systems. For paired-end sequencing, two FASTQ files (R1 and R2) are generated. Raw data in FASTQ format is ready for subsequent quality control and alignment.

3.6 Bioinformatic Analyses

1. Quality assessment. The quality of raw reads is determined by *FastQC* [6]. There are four main metrics to be considered for RNA-seq quality control: the sequencing depth (which affects statistical power), the per-base quality (this score is expected to gradually decline towards the end of the read), the GC content and the adapter contamination (for good quality sequencing, the first half of the read must be adapter-free; see **step 2**). Each sample has to be processed individually. *FastQC* can be easily run as follows:

```
fastqc sampleName_R1.fastq.gz sampleName_R2.fastq.gz
```

Results are reported in HTML format that can be opened in any web explorer for visualization.

2. Adapter trimming. Sequencing adapters can be removed with tools as *trimmomatic* (*see* **Note 5**). However, read trimming may introduce unwanted source of bias. It is recommended to run through the analysis with and without trimming to check the effect of read trimming on the final results (e.g., total number of genes detected, list of differentially expressed

genes). If trimming is needed, use a minimum read size (MINLEN parameter) of at least 36 bp [22]. Note that for PE sequencing data, four output files have to be specified: 2 for the paired output were both read within a pair remained after the trimming, and 2 for the unpaired output where only one read within the pair survived the trimming. Each sample has to be processed individually. Therefore, the basic command to run *trimmomatic* is the following:

```
java -jar path/to/trimmomatic-0.39.jar \
PE sampleName_R1.fastq.gz sampleName_R2.fastq.gz \
sampleName_R1_trimmomatic.fastq.gz sampleName_R1_trimmomati-
c_unpaired.fastq.gz sampleName_R2_trimmomatic.fastq.gz sam-
pleName_R2_trimmomatic_unpaired.fastq.gz \
ILLUMINACLIP:path/to/Illumina_adapters.fa:2:30:10 \
MINLEN:36 \
```

3. Prepare reference genome indexes. Before alignment, genome indexes have to be created in order to use the reference genome for RNA-seq. This step is performed with the genomeGenerate mode of *STAR* [9] as follows (*see* **Note 6**):

```
STAR --runMode genomeGenerate \
--genomeDir path/to/genomeDirectory \
--genomeFastaFiles Mus_musculus.GRCm39.dna.primary_assembly.
fa \
--sjdbGTFfile Mus_musculus.GRCm39.111.gtf \
--sjdbOverhang 99 \
```

Genome indexes will be created inside the path/to/genomeDirectory folder. Files within this folder should not be modified.

4. Read alignment to the reference genome. The resulting reads are aligned to the reference genome with *STAR* (Spliced Transcripts Alignment to a Reference). Although memory intensive, STAR outperforms other aligners in terms of mapping speed, alignment precision, and sensitivity [9]. Each sample has to be aligned individually. To obtain an aligned BAM file sorted by genomic coordinates as output, the basic command to run STAR is as follows (*see* **Note 7**):

```
STAR --runThreadN 8 \
--genomeDir path/to/genomeDirectory \
--readFilesIn sampleName_R1.fastq.gz sampleName_R2.fastq.gz \
--outSAMtype BAM SortedByCoordinate \
--chimSegmentMin 20 \
```

```
--chimOutType WithinBAM
```

The file `Aligned.sortedByCoord.out.bam` will be generated.

5. Quantification of transcript levels. Transcript abundance quantification can be performed with *HTseq* (htseq-count) [10]. This step involves counting the reads that are assigned to a list of genomic features (i.e., genes and other transcripts such as micro-RNAs or long-non-coding-RNAs). *HTseq* has three different modes to deal with reads that overlap with more than one feature (`union`, `intersection_strict`, `intersection_nonempty`), the mode `union` is recommended for most use cases. *HTseq* needs an alignment SAM file (obtained in the previous step with STAR) and a file containing the genomic features in GTF format (the same GFT file used in **step 3**). First, we need to convert our alignment BAM file into SAM format with *samtools* [8]:

```
samtools view -h -o Aligned.sortedByCoord.out.sam Aligned.
sortedByCoord.out.bam \
```

Then, transcript quantification can be performed (*see* **Note 8**):

```
htseq-count -c htseq_sampleName.tsv -m union --stranded=no -r
pos Aligned.sortedByCoord.out.sam Mus_musculus.GRCm39.111.gtf
\
```

The output of this command will be a two-column file in TSV format, with the first column containing the genes ENSEMBL_ID and the second column the read count assigned to each gene.

6. Load count data into R for downstream analyses. The commands presented in the following steps must be run within the R environment. The Bioconductor package *DESeq2* has a function called `DESeqDataSetFromHTSeqCount` that allows a straightforward solution to load *HTseq* count data. The user has to specify the directory containing the count files in TSV format and an array of sample conditions (i.e., if the sample belongs to a TR-deficient or wild-type mouse). The basic code to load count data with the *DESeq2* package is the following:

```
library(DESeq2)
directory <- "/path/to/your/countFiles/"
sampleFiles <- list.files(directory)
sampleTable <- data.frame(sampleName=sampleFiles, fileName=-
sampleFiles, condition=factor(sampleCondition, levels=c("WT",
```

```
"TRdeficient"))
```

Then, the main object of the *DESeq2* workflow (`DESeqDataSet` object) is built as shown:

```
ddsHTSeq <- DESeqDataSetFromHTSeqCount(sampleTable=sampleTa-
ble, directory = directory, design= ~ condition)
```

7. Quality control of gene expression data. The basic quality control of gene expression data can be performed with the *RNAseqQC* R package [14], which includes functions for the evaluation of several metrics, such as library complexity, total counts, and the number of detected genes in each sample. Also, Principal Component Analysis (PCA) or unsupervised hierarchical clustering based on the Euclidean Distance can be performed with the *RNAseqQC* package to detect the presence of outliers and explore the dataset clustering patterns (*see* **Note 9**).

8. Differential expression analysis. The gene expression profiles in TR-deficient mice can be obtained through gene expression analysis. All the steps of the differential expression analysis are included in the `DESeq` function from *DESeq2*. It is important to note that *DESeq2* uses a built-in normalization method that corrects for library size [13]. Hence, raw count data must be used as input for differential expression analysis with this package. Either the Wald test (default) of the likely ratio test can be used to calculate the p-values and the fitting of the log2FoldChange (LFC). Therefore, differential expression analysis between TR-deficient and wild-type (WT) mice is carried out as follows:

```
DEA <- DESeq(ddsHTSeq)
DEA.res <- results(DEA.res, pAdjustMethod='BH")
head(DEA.res)
```

The resulting `DEA.res` object is a table which contains, among other statistics, the LFC with respect to the TR-deficient mice (meaning that a positive LFC in a specific gene indicates a higher expression of such gene in TR-deficient mice compared to the WT, while a negative LFC will indicate a lower expression of a in TR-deficient mice compared to the WT). The nominal p-value and the Bonferroni-Hochberg adjusted p-value are also associated to each gene (*see* **Note 10**). A volcano plot, which shows the LFC of each gene on the X-axis with their corresponding -$\log_{10}$(adjusted p-value) on the Y-axis, offers an overview of the result of differential expression analysis.

9. Count data normalization. Although *DESeq2* employs un-normalized data as input for differential expression analysis, normalized count data such as counts per million (CPM) or fragments per kilobase per million mapped fragments (FPKM) are needed for the visualization of gene expression levels in plots such as Heatmaps. The functions `fpkm` from the *DESeq2* package and the function `cpm` from the *edgeR* package [15] are useful for returning FPKM and CPM normalized counts, respectively.

10. Map between feature identifiers. The obtention of different gene identifiers can be performed using the *org.Mm.eg.db* Bioconductor library, which contains genome-wide annotation for mouse [12]. The obtention of the gene symbol providing the ENSEMBL ID (that is the default *HTseq* output) can be easily achieved as follows:

```
library(org.Mm.eg.db)
mapIds(org.Mm.eg.db, rownames(ddHTseq), keytype="ENSEMBL",
column="SYMBOL")
```

11. Functional enrichment. Gene Set Enrichment Analysis (GSEA) [23] provides a functional interpretation of gene expression patterns found in TR-deficient mice. The `gseKEGG` function from the *clusterProfiler* R package [16] can be used to perform an enrichment analysis over the regulatory and metabolic pathways from the KEGG database. The input of the GSEA is the complete list of genes ranked according to the result of differential expression, either the p-value or the LFC or a score that combines both such as multiplying the -log10(p-value) by the sign (±1) of the LFC. Then GSEA is performed on the mice KEGG pathways with the *clusterProfiler* package:

```
results.gsekegg <- gseKEGG(geneList = genes, organism =
"mmu", keyType = "ncbi-geneid", minGSSize = 3, maxGSSize =
500, pvalueCutoff = 0.1, pAdjustMethod = "BH")
```

The *enrichplot* package [17] is mainly designed to work with clusterProfiler package results, and implements a variety of visualization methods such as `gseaplot` or `ridgeplot` for interpreting functional enrichment results.

4 Notes

1. Depending on the RNA extraction kit used, there may be some reagents or equipment that have to be supplied by the user such as sterile low-binding RNAase-free tubes, sterile RNase-free pipette tips, 96–100% ethanol, microcentrifuge, vortexer or heat block.
2. Although tissues can be preserved indefinitely in RNAlater® at −20 °C, −80 °C is recommended for long-term optimal RNA preservation.
3. When working with precious or limited samples, it is recommended to perform an extra elution to ensure you recover the maximum RNA from the spin column membrane.
4. The Qubit RNA HS (high sensitivity) assay can determine accurate sample concentrations from 0.2 to 200 ng/μL (assay range of 4–200 ng). While the Qubit BR (broad range) assay is suitable for samples ranging 0.5–1200 ng/μL (assay range of 10–1200 ng).
5. After the trimming, repeat the *FastQC* check to ensure the quality of the processed reads.
6. The parameter `--sjdbOverhang` has to be adjusted to 1 bp less than your read size. In the example it was set to 99 assuming that 100 bp PE sequencing was performed.
7. It is possible to detect gene fusions from RNA-seq data. A chimeric alignment has two segments that are far from each other that belong to different strands or to different chromosomes. Chimeric alignments can be included in the BAM file together with normal alignment by setting the parameters `--chimSegmentMin` and `--chimOutType`. For 100 bp PE reads, if you set `--chimOutType 20`, chimeric alignments with 60 bp in one chromosome and 20 bp in another will be included, while one with 65 + 15 bp will be discarded. The resulting BAM file can be used for both transcript quantification with tools such as *HTseq* [10] and to detect gene fusions.
8. The default HTseq `--stranded` parameter is `yes`, therefore if your data was generated with an unstranded library (as the Illumina TruSeq RNA Library Prep Kit v2) make sure to set `--stranded = no`, otherwise half of the reads will not be assigned.
9. Although the filtering of low-count genes is not mandatory before the differential expression analysis with *DESeq2*, it may reduce memory usage and increase the speed of modeling during the differential expression analysis. Removing features with less than 10 counts is reasonable for most bulk RNA-seq datasets. Note this filtering can be skipped.

10. The user can select a different p-value adjustment method with the parameter `pAdjustMethod` of the `results` function.

References

1. Conesa A, Madrigal P, Tarazona S, Gomez-Cabrero D, Cervera A, McPherson A et al (2016) A survey of best practices for RNA-seq data analysis. Genome Biol 17(1):1–19
2. Stark R, Grzelak M, Hadfield J (2019) RNA sequencing: the teenage years. Nat Rev Genet 20(11):631–656
3. Thind AS, Monga I, Thakur PK, Kumari P, Dindhoria K, Krzak M et al (2021) Demystifying emerging bulk RNA-Seq applications: the application and utility of bioinformatic methodology. Brief Bioinform 22(6):bbab259
4. Simoneau J, Dumontier S, Gosselin R, Scott MS (2021) Current RNA-seq methodology reporting limits reproducibility. Brief Bioinform 22(1):140–145
5. Ruiz-Llorente L, Contreras-Jurado C, Martínez-Fernández M, Paramio JM, Aranda A (2018) Thyroid hormone receptors regulate the expression of microRNAs with key roles in skin homeostasis. Thyroid 28(7):921–932
6. Andrews S (2010) FastQC A quality control tool for high throughput sequence data. Available from: https://www.bioinformatics.babraham.ac.uk/projects/fastqc/. Accessed 3 Mar 2024
7. Bolger AM, Lohse M, Usadel B (2014) Trimmomatic: a flexible trimmer for Illumina sequence data. Bioinformatics 30(15):2114–2120
8. Li H, Handsaker B, Wysoker A, Fennell T, Ruan J, Homer N et al (2009) The sequence alignment/map format and SAMtools. Bioinformatics 25(16):2078–2079
9. Dobin A, Davis CA, Schlesinger F, Drenkow J, Zaleski C, Jha S et al (2013) STAR: ultrafast universal RNA-seq aligner. Bioinformatics 29(1):15–21
10. Anders S, Pyl PT, Huber W (2015) HTSeq—a python framework to work with high-throughput sequencing data. Bioinformatics 31(2):166–169
11. Gentleman RC, Carey VJ, Bates DM, Bolstad B, Dettling M, Dudoit S et al (2004) Bioconductor: open software development for computational biology and bioinformatics. Genome Biol 5(10):R80
12. Carlson M (2019) org.Hs.eg.db: genome wide annotation for human. R package version 382. Available from https://bioconductor.org/packages/release/data/annotation/html/org.Hs.eg.db.html
13. Love MI, Huber W, Anders S (2014) Moderated estimation of fold change and dispersion for RNA-seq data with DESeq2. Genome Biol 15(12):550
14. Ziebell F (2022) RNAseqQC: quality control for RNA-Seq data. R package version 0.1.4. Available from https://cran.r-project.org/web/packages/RNAseqQC/index.html
15. Smyth G (2009) edgeR: a Bioconductor package for differential expression analysis of digital gene expression data. Available from: https://www.ncbi.nlm.nih.gov/pmc/articles/PMC2796818/. Accessed 3 Mar 2024
16. Yu G, Wang LG, Han Y, He QY (2012) clusterProfiler: an R package for comparing biological themes among gene clusters. OMICS 16(5):284–287
17. Yu G (2023) Enrichplot: visualization of functional enrichment result. R package version 1.23.1. Available from https://bioconductor.org/packages/release/bioc/html/enrichplot.html
18. Gallego Romero I, Pai AA, Tung J, Gilad Y (2014) RNA-seq: impact of RNA degradation on transcript quantification. BMC Biol 12(1):42
19. Schroeder A, Mueller O, Stocker S, Salowsky R, Leiber M, Gassmann M et al (2006) The RIN: an RNA integrity number for assigning integrity values to RNA measurements. BMC Molecular Biol 7(1):3
20. Cieslik M, Chugh R, Wu YM, Wu M, Brennan C, Lonigro R et al (2015) The use of exome capture RNA-seq for highly degraded RNA with application to clinical cancer sequencing. Genome Res 25(9):1372–1381
21. Illumina. "Evaluating RNA quality from FFPE samples" Illumina publication number 470-2014-001
22. Williams CR, Baccarella A, Parrish JZ, Kim CC (2016) Trimming of sequence reads alters RNA-Seq gene expression estimates. BMC Bioinformatics 17(1):103
23. Subramanian A, Tamayo P, Mootha VK, Mukherjee S, Ebert BL, Gillette MA et al (2005) Gene set enrichment analysis: a knowledge-based approach for interpreting genome-wide expression profiles. Proc Natl Acad Sci (PNAS) 102(43):15545–15550

Chapter 11

Validation of miR-21, miR-31, miR-34A, and miR-203 Changes by Quantitative Polymerase Chain Reaction in Isolated Keratinocytes

Constanza Contreras-Jurado

Abstract

MiRNAs play integral roles in diverse cellular functions, and their dysregulation is central to various pathological processes. Thyroid hormone (TH) is indispensable for numerous physiological processes, and it has been shown that multiple miRNAs regulate TH signaling in various tissues.

This chapter describes a method for validating changes in miRNA observed through RNAseq. The method involves measuring the expression of mature miRNA levels in keratinocytes using the stem-loop real-time PCR. This method utilizes stem-loop RT primers to create cDNA for specific miRNAs in a single RT reaction. The keratinocytes used are derived from the total skin of four-day-old wild-type controls and double thyroid receptors knock-out mice (dTRKO).

Key words Thyroid hormone receptors (TR), Epidermal skin, microRNAs, Keratinocytes, Retrotranscription, Stem-loop Real-time PCR, Double thyroid receptors Knock-Out mice

1 Introduction

MicroRNAs (miRNAs) are small endogenous non-coding RNAs, ranging from 19 to 25 nucleotides in size, which exert post-transcriptional regulation over gene expression. A single miRNA can target numerous mRNAsr, thereby influencing the expression of multiple genes often associated with interconnected functional pathways. Human cells harbor over 2300 distinct miRNAs, exhibiting expression patterns that vary with time and tissue type [1].

Most human miRNA sequences are transcribed within intronic regions of both coding and non-coding transcripts, although a minority originate from exonic regions. These miRNA genes are typically transcribed by either polymerase II or III, yielding primary miRNAs (pri-miRNAs), which may encompass sequences for multiple miRNAs and extend over hundreds of nucleotides. Subsequent processing involves cleavage by the Drosha-DGCR8

Constanza Contreras-Jurado (ed.), *Thyroid Hormones: Methods and Protocols*, Methods in Molecular Biology, vol. 2876, https://doi.org/10.1007/978-1-0716-4252-8_11,

complex, yielding hairpin-shaped stem-loop structures known as precursor miRNAs (pre-miRNAs), typically around 70 nucleotides long. Exportin 5 facilitates the transportation of pre-miRNAs out of the nucleus. Further cytoplasmic processing by Dicer1-TARBP2, an RNase III enzyme, results in the generation of a duplex comprising miRNA and miRNA strands, typically 18–25 nucleotides in length, with one strand designated as the guide strand. The guide strand integrates into the RNA-induced silencing complex (RISC), a multiprotein ribonucleoprotein complex, which orchestrates the modulation of target gene transcription. Alternative pathways, independent of Drosha-DGCR8 or Dicer, have also been described [2, 3].

MiRNAs play integral roles in diverse cellular functions, and their dysregulation is central to various pathological processes, making them promising candidates for biomarkers for diagnosing and predicting diseases because of their stability in human fluids. A single miRNA can control entire cellular pathways by interacting with a wide range of target genes. This makes miRNAs very promising as therapeutic tools for restoring normal cell functions that are disrupted in disease [4, 5].

Thyroid hormone (TH) is indispensable for numerous physiological processes across vertebrate tissues, and aberrant TH signaling underlies several human diseases. TH actions are predominantly mediated by thyroid hormone receptors (TRs), members of the nuclear receptor superfamily functioning as ligand-dependent transcription factors. While the thyroid gland primarily secretes thyroxine (T4), the more biologically active hormone is triiodothyronine (T3), with higher receptor affinity. T3 availability for receptor binding is regulated by cell-specific expression of deiodinases (DIOs), a family of selenoenzymes. DIO1 and DIO2 catalyze the conversion of T4 to T3 in target tissues, augmenting intracellular levels of the active hormone, whereas DIO3 inactivates hormones by converting T4 and T3 to the inactive metabolites reverse T3 (rT3) and T2, respectively [6].

Multiple miRNAs target deiodinases and TRs, thereby regulating TH signaling in various tissues. Conversely, THs have been shown to modulate the expression of specific miRNAs and their mRNA targets across different cell types and organs. In our laboratory, RNAseq analysis was employed to compare the skin miRnome of wild-type mice with genetically modified mice lacking both TRα1 and TRβ, the primary thyroid hormone binding isoforms. Alterations in key miRNAs crucial for skin physiopathology were validated via stem-loop quantitative polymerase chain reaction, and the expression levels of their target mRNAs were assessed using real-time polymerase chain reaction [7].

In our laboratory, we have shown that the skin of TRα1/TRβ knockout mice exhibited dysregulated levels of >50 miRNAs. Notably, among the downregulated species were several miRNAs, including miR-21, miR-31, miR-34, and miR-203, known for their pivotal roles in maintaining skin homeostasis [8].

Quantitative reverse transcriptase polymerase chain reaction (RT-qPCR) is widely considered the gold standard for quantifying miRNAs due to its optimal balance of cost, precision, sample size, and broad functional dynamic range [26]. In this method, extracted RNA is converted into cDNA through reverse transcription using universal primers, miRNA-specific primers, or miRNA-specific stem-loop primers. The mature miRNA sequences present in pre-miRNAs and pri-miRNAs pose challenges for designing specific and sensitive primers; however, stem-loop primers can effectively differentiate between various miRNA forms [27], thus enhancing specificity and reproducibility [28]. The resulting cDNA is then subjected to qPCR, which can be performed using a miRNA-specific forward primer, a universal reverse primer, and SYBRGreen, or TaqMan probes. RT-qPCR's primary advantage is its capacity for high specificity and quantitative accuracy.

Here, we detail protocols for miRNA isolation from keratinocytes derived from the total skin of four days old wild-type, confirmation of miRNA changes using stem-loop quantitative polymerase chain reaction, and evaluation of the expression levels of their target mRNAs via real-time polymerase chain reaction.

2 Materials

2.1 Animals

C57B/6 wild-type mice and $TR\alpha^{-/-}/TR\beta^{-/-}$ double KO mice (with a deletion of TRα1 but not TRα2 and lack all TRβ isoforms) are bred and maintained in a specific pathogen-free (SPF) environment according to animal facility regulations.

Neonates are used within 4 days of birth (postnatal days 0 ~ 4) and either female or male mice can be used.

2.2 Skin Digestion

1. PBS containing antibiotic/antimycotic, pH 7.4.
2. 35-mm sterile Petri dish.
3. Trypsin solution 0,25%.

2.3 Isolation of Keratinocytes from Skin Epidermis

1. Eagle's minimum essential medium (EMEM): Earle's Balanced Salt Solution, non-essential amino acids, 2 mM L-glutamine, 1 mM sodium pyruvate, and 1500 mg/L sodium bicarbonate.
2. 4% fetal bovine serum (FBS) pretreated with Chelex 100 resin: 180 g Chelex 100 resin, 500 mL water and 500 mL Bovine serum, pH 7.
3. Plating medium: EMEN, 0.2 mM $CaCl_2$,4%, FBS 4%, antibiotic-antimycotic 1%.
4. 100 μm filter.

2.4 Quantitative Real-Time PCR Assays from miRNA

2.4.1 Purification of Total RNA Including miRNA

1. Microcentrifuge.
2. Filtered pipette tips.
3. Vortex.
4. RNA/DNAse-free1.5mLtubes.
5. 1X PBS.
6. TRIZOL.
7. Chloroform, stored at room temperature.
8. Isopropanol, stored at room temperature.
9. 75% Ethanol, stored at −20 °C.
10. Ice.
11. Nuclease-free water.

2.4.2 RNA Quality and Concentration Determinations

1. NanoDrop Spectrophotometer.
2. Agarose gel electrophoresis.
3. UV transilluminator.
4. Electrophoresis system.
5. RNA markers.
6. 28S and 18S eukaryotic ribosomal RNAs.
7. FA gel buffer 10X: 10 mM EDTA, 200 mM MOPS, 50 mM sodium acetate (NaOAc) and adjust volume with DEPC-treated deionized water. Rectify pH to 7.0 and store in the dark.
8. Running buffer 1X: Dilute FA gel buffer 10X o 1X with DEPC-treated deionized water, to a final 2% formaldehyde solution.
9. RNA loading buffer 5X: 40% 10X gel buffer, 4 mM EDTA, 7.2% formaldehyde solution, 30.84% formamide, 20% glycerol, 0.16% saturated bromophenol blue solution. Store at 4 °C.
10. 1.2% denaturing agarose gel: Add 1.2% agarose (w/v) in an appropriate volume of 1X FA gel buffer. Completely melt the agarose in a microwave oven, cool it to about 65 °C, and then add a final concentration of 1.8% formaldehyde solution and 50 ng/mL ethidium bromide. Mix the solution and pour it onto the gel support to solidify. Finally, equilibrate the gel in 1x running buffer for at least 30 min.
11. Bioanalyzer (optative).

2.4.3 Reverse Transcription

1. Thermal cycler.
2. RNA/DNAse-free 0.2 mL tubes.
3. Filtered pipette tips.
4. Nuclease-free water.

5. DNAse and Reaction Buffer, stored at −20 °C.
6. 25 mM EDTA.
7. TaqMan MicroRNA Reverse Transcription Kit (Applied Biosystems, cat. #4366597), stored at −20 °C.
8. Specific RT primers hsa-miR-21 mouse TaqMan MicroRNA primer (Applied Biosystems, cat. #000397).
1. Specific RT primers mmu-miR-31 mouse TaqMan MicroRNA primer (Applied Biosystems, cat. #000185).
2. Specific RT primers hsa-miR-34a mouse TaqMan MicroRNA primer (Applied Biosystems, cat. #000426).
3. Specific RT primers U6 snRNA TaqMan MicroRNA primer (Applied Biosystems, cat. #001973) for normalization.

2.4.4 miRNA PCR Amplification

1. Real-time PCR machine.
2. RNA/DNAse-free 1.5 mL tubes.
3. Filtered pipette tips.
4. 96-well optical reaction plate.
5. Optical adhesive cover.
6. Nuclease-free water.
7. TaqMan Universal PCR Master Mix No AmpERASE (Applied Biosystems 4,324,018).
1. Specific Forward primers hsa-miR-21 mouse TaqMan MicroRNA primer (Applied Biosystems, cat. #000397).
2. Specific Forward primers mmu-miR-31 mouse TaqMan MicroRNA primer (Applied Biosystems, cat. #000185).
3. Specific Forward primers hsa-miR-34a mouse TaqMan MicroRNA primer (Applied Biosystems, cat. #000426).
4. Specific Forward primers U6 snRNA TaqMan MicroRNA primer (Applied Biosystems, cat. #001973) for normalization.

3 Methods

Keratinocytes are derived from the total skin of four days old wild-type controls and double KO mice. Cell culture is carried out in a class II laminar flow Hood and incubations are at 37 °C in a humidified atmosphere of 5% CO2.

3.1 Skin Digestion

1. Euthanize the postnatal day 4 old C57BL/6 Wt and TR$\alpha^{-/-}$/TRb$^{-/-}$ double KO mice neonatal pups with CO_2.
2. Clean the mice with water and then with 70% ethanol.

3. To peel neonatal skin off the body, first cut off limbs just above the wrist and joints and cut off the tail from the base and discard. Make a longitudinal incision on the dorsum from the base of the tail to the snout. Remove the skin using sterile forceps, taking care not to break viscera or drag fat or muscles (*see* **Note 1**).
4. Rinse the peeled neonatal skin with sterile PBS containing antibiotic/antimycotic.
5. Transfer the skin to a 35-mm sterile Petri dish. Stretch the skin well so that the skin is fully extended, epidermis side up and dermis down. Add 2 mL of cold trypsin solution 0.25%. Avoid trypsin contact with epidermis (*see* **Note 2**).
6. Digest the skin 18 h at 4 °C in a refrigerator.

3.2 Isolation of Keratinocytes from Skin Epidermis

1. Next day rinse the peeled neonatal skin with sterile PBS to wash away excess dispase.
2. Transfer the skin to a new 35-mm sterile Petri dish. Stretch the skin well, dermis down.
3. To peel epidermis off the dermis, use one forceps to gently lift a corner of the skin off and the other forceps to carefully peel dorsal skin off with one continuous motion (*see* **Note 3**).
4. Transfer the epidermis, from 1 to 4 animals to 35-mm sterile Petri dish with 1.5 mL of plating medium and cut the epidermis into pieces.
5. Transfer the pieces with plating medium to 50 mL tubes. Add another 2 mL of plating medium to the tube and carefully resuspend by pipetting the cell suspension up and down a few times to disperse cell clumps using a serological pipette (*see* **Note 4**).
6. Place a 100 μm filter in a clean tube with a capacity of 50 mL and filter the medium with the disaggregated cells through (see **Note 5**).
7. Centrifuge the filtered cells at 180 g for 5 min. Aspirate the supernatant and resuspend the cell pellet by making pools of at least 3 animals.

3.3 Quantitative Real-Time PCR Assays from miRNA

RNA extraction can be done by guanidinium thiocyanate-phenol-chloroform protocol, but we strongly recommend using a column-based RNA extraction kit (*see* **Note 6**).

3.3.1 Purification of Total RNA Including miRNA

1. Homogenization: Transfer cells to a microcentrifuge tube, add 1 mL of guanidinium thiocyanate per 5–10 × 10^6 cells and resuspend the mixture well. Store homogenate for 5 min at room temperature (*see* **Note 7**).

2. RNA extraction: Add 0.2 mL of chloroform and mix vigorously. Store sample 5–10 min at room temperature. After this incubation time, centrifuge mix at 12000 g for 15 min at room temperature (*see* **Note 8**).
3. RNA precipitation: Transfer aqueous phase into a new microcentrifuge tube, add 0.5 mL of isopropanol and stir strongly. Incubate 10 min at room temperature. After that centrifuge mix at 12000 g for 8 min at 4 °C. Eliminate the supernatant.
4. RNA wash: Mix RNA pellet with 1 mL 75% cold ethanol. Centrifuge mix at 7500 g for 5 min at 4 °C. Eliminate the supernatant.
5. Solubilization: Air dry RNA pellet for 5–10 min on ice. Dissolve RNA in 20 μL Nuclease-free water.

3.3.2 RNA Quality and Concentration Determinations

1. Determine the absorbance and concentration (ng/μL) for each RNA sample with a NanoDrop-1000 spectrophotometer. The ratio of the absorbance at 260 and 280 must be greater than 1.8 but a value of 2 is preferred (*see* **Note 9**).
2. Analyze RNA integrity by denaturing agarose gel electrophoresis 1 μg RNA with 5X RNA loading buffer and electrophorese in a 1.2% denaturing agarose gel at 100 V. Visualized gel under the UV transilluminator, two distinctive bands of about 1.9 and 4.7 kb for mouse RNA should be observed. The ratio of 28S to 18S eukaryotic ribosomal RNAs should be approximately 2:1. This ratio indicates that no significant degradation of RNA has occurred. When RNA samples have degraded, this ratio will be reversed, as the 28S ribosomal RNA is typically degraded to an 18S-like species (*see* **Note 10**).

3.3.3 Reverse Transcription

1. Dilute aliquots of all RNA samples with Nuclease-free water to yield 1 μg/μL solutions of total RNA.
2. Treat the RNA samples with DNAse for digestion of single and double-stranded DNA. 1 μL RNA (at a concentration 1 μg/μL) and 1 μL of buffer, 1 μL DNAse and, 7 μL Nuclease-free water. Incubate 15 min at room temperature. Inactivate DNAse with 1 μL of 25 mM EDTA and incubate for 10 min at 65 °C.
3. We strongly recommend using a specific kit that only quantifies mature miRNA, not precursors. In our laboratory, we use a highly sensitive Taq Man MicroRNA reverse transcript kit that requires only 1 to 10 nanograms of total RNA.

   In a polypropylene tube, prepare the mix on ice: 0.05 μL 100 mM dNTPs with dTTP, 0.33 μL Multiscribe reverse transcriptase 50 U/ μL, 0.5 μL 10X reverse transcription buffer, 0.068 μL RNAse inhibitor 20 U/μL, 2.06 μL Nuclease-free water, mix gently.

4. To this 3 μL mix add 1 μL RNA at 10–15 ng/ μL and 1 μL specific forward RT primers (miR-21, miR-31, miR-34A, miR-203, and U6 separately) (*see* **Note 11**).
5. Incubate in a thermal cycler: 30 min 16 °C, 30 min 42 °C, 5 min 85 °C and leave it at 4 °C until you are ready to load the 96-well plate to PCR amplification.

3.3.4 PCR Amplification

miRNAs must be quantified minimum in triplicate using TaqMan Universal PCR Master Mix No AmpERASE (Applied Biosystems, Foster City, CA).

1. Add 10.15 μL of Nuclease-free water to each RT reaction.
2. Use the stock cDNA from the stem-loop RT reactions to generate a standard curve for miR-21, miR-31, miR-34A, miR-203, and U6 qRT-PCR reactions (for Normalization of Stem-Loop qRT-PCR results) (*see* **Note 12**).
3. For amplification and detection of mature-specific miRNA, in a 96-well optical reaction plate, perform triplicate qRT-PCR reactions (10 μL reaction volumes) for standards, samples, and negative controls using the miRNA-specific TaqMan probes provided in the MicroRNA Assay Kit. Reaction: 5 μL TaqMan 2X PCR Master Mix, 2 μL specific-cDNA RT, 0.5 μL specific miR-primer, and 2.5 μL Nuclease-free water.
4. Seal the plate with an optical adhesive cover and centrifuge the plate briefly to spin down the content and eliminate any air bubbles.
5. Load the plate into a sequence detector and run the reaction with the following program:

   Stage 1 (TaqMan hot start): 95 °C for 10 min.

   Stage 2 (amplification): 95 °C for 15 s then 60 °C for 1 min (37 cycles).

3.3.5 Normalization and Analysis Methods

Data analysis is done using the comparative quantification $2^{\Delta\Delta Ct}$ method, and data is corrected with the U6 small RNA used as an internal control.

1. Calculating fold change using the equation:

$$\text{Ratio}: \frac{2^{Ct}_{BGOI}\ {}^{-Ct}_{AGOI}}{2^{Ct}_{BRAF}\ {}^{-Ct}_{AREF}} = \frac{2^{\Delta Ct}_{GOI}}{2^{\Delta Ct}_{REF}} = 2^{\Delta\Delta Ct}$$

   Ct: number of cycles required to obtain threshold level.

   B: sample B.

   A: sample A.

   GOI: gene of interest.

   REF: reference or housekeeping gene.

2. The reactions with cycle threshold (CT) values higher than 34 and three or more melting peaks must be excluded.
3. Ct value of each miRNA is normalized using the U6 small RNA.
4. The ΔCt method is used to calculate the relative expression of the miRNAs individually [9].
5. The ΔCt corresponds to the difference between Ct from miRNA target and Ct of the endogenous miRNA (U6). Data are transformed by 2-ΔCt method and used in the graphics.

4 Notes

1. Carefully peel the whole skin off the body in one continuous motion as one piece (careful not to break skin into pieces as this will result in cell loss during the dispase digestion step).
2. Add the trypsin by resting the pipette on the bottom of the plate, away from the skin, so that the skin raises the level of trypsin.
3. Epidermal sheet is much thinner (semitransparent) than the dermis (opaque).
4. Swab the plate with 500 μL of plating medium to try to catch all the debris.
5. Cells can be frozen in a special freezing medium after being filtered.
6. Due to the very small size of miRNAs, we strongly recommend using a column-based RNA extraction kit to reach the optimum yield of miRNA.
7. The mixture can be stored at −20 °C until it is processed.
8. This step can be repeated if very clean RNA is needed but it also reduces the yield.
9. The 260/280 with a lower value ratio of 1.8–2.0 suggests high levels of protein within a sample, while a 260/230 ratio below 1.8 can indicate the presence of organic contaminants.
10. RNA can be stored at −80 °C until it is processed.
11. Primer's forward sequences can also be obtained from mirBase (https://www.mirbase.org). To design the forward primers, it is necessary to replace the uracil bases with thymine bases in the mature miRNA sequence before ordering the oligo.
12. For Normalization of Stem-Loop qRT-PCR results can be used as well TaqMan Rodent GAPDH Control Reagents.

References

1. Lee Y, Kim M, Han J et al (2004) MicroRNA genes are transcribed by RNA polymerase II. EMBO J 23(20):4051–4060. https://doi.org/10.1038/sj.emboj.7600385. Epub 2004 Sep 16. PMID: 15372072; PMCID: PMC524334
2. Bartel DP (2004) MicroRNAs: genomics, biogenesis, mechanism, and function. Cell 116(2): 281–297. https://doi.org/10.1016/s0092-8674(04)00045-5
3. Michlewski G, Cáceres JF (2019) Post-transcriptional control of miRNA biogenesis. RNA 25(1):1–16. https://doi.org/10.1261/rna.068692.118. Epub 2018 Oct 17. PMID: 30333195; PMCID: PMC6298569
4. Ho PTB, Clark IM, Le LTT (2022) MicroRNA-based diagnosis and therapy. Int J Mol Sci 23(13):7167. https://doi.org/10.3390/ijms23137167. PMID: 35806173; PMCID: PMC9266664
5. Diener C, Keller A, Meese E (2022) Emerging concepts of miRNA therapeutics: from cells to clinic. Trends Genet 38(6):613–626. https://doi.org/10.1016/j.tig.2022.02.006. Epub 2022 Mar 15
6. Lasa M, Contreras-Jurado C (2022) Thyroid hormones act as modulators of inflammation through their nuclear receptors. Front Endocrinol 13:937099. https://doi.org/10.3389/fendo.2022.937099. PMID: 36004343; PMCID: PMC9393327
7. Aranda A (2021) MicroRNAs and thyroid hormone action. Mol Cell Endocrinol 525:111175. https://doi.org/10.1016/j.mce.2021.111175. Epub 2021 Jan 27
8. Ruiz-Llorente L, Contreras-Jurado C, Martínez-Fernández M et al (2018) Thyroid hormone receptors regulate the expression of microRNAs with key roles in skin homeostasis. Thyroid 28(7):921–932. https://doi.org/10.1089/thy.2017.0369. Epub 2018 Jun 11
9. Schmittgen T, Livak K (2008) Analyzing real-time PCR data by the comparative C_T method. Nat Protoc 3:1101–1108. https://doi.org/10.1038/nprot.2008.73

Part IV

Thyroid Hormone Transporter MCT8

Chapter 12

Magnetic Resonance Imaging Techniques for Investigating the MCT8-Deficient Brain in Murine Disease Models

Marina Guillén-Yunta, Ana Guadaño-Ferraz, and Víctor Valcárcel-Hernández

Abstract

Magnetic resonance imaging (MRI) techniques have emerged as powerful tools for unraveling the pathophysiology of rare diseases, mainly due to their pivotal role in early diagnosis, disease characterization, and treatment monitoring in a non-invasive manner. In this chapter, we will review two essential MRI tools used for studying and evaluating the pathophysiology of Allan-Herndon-Dudley Syndrome or MCT8 deficiency, a rare disease caused by inactivating mutations in the *SLC16A2* gene, encoding for the thyroid hormone-specific transmembrane transporter MCT8. These two MRI techniques are time-of-flight magnetic resonance angiography (TOF-MRA) and diffusion tensor imaging (DTI).

Key words MRI, MRA, Angiography, Diffusion tensor imaging, MCT8, AHDS, Brain, Myelin, Blood-brain barrier, Axonal damage

1 Introduction

While many brain pathological alterations related to thyroid hormones, such as the ones linked to monocarboxylate transporter 8 (MCT8) deficiency, have been well defined using postmortem histology, electrophysiology, or molecular biology techniques among others, the ability to non-invasively evaluate longitudinal progression across various conditions is critical to understanding (i) the evolution of different pathologies and (ii) the potential outcomes of therapeutic approaches.

Magnetic resonance imaging (MRI) parameters have been consistently used in clinical research in different pathologies, such as Alzheimer's disease [1, 2], multiple sclerosis [3, 4], and MCT8 deficiency itself [5–7]. MRI analyses have also been used relatively recently in preclinical research involving neurodegenerative and autoimmune diseases [8–10], as well as in rare neurodevelopmental disorders like MCT8 deficiency [11, 12]. However, the utility of

Constanza Contreras-Jurado (ed.), *Thyroid Hormones: Methods and Protocols*, Methods in Molecular Biology, vol. 2876, https://doi.org/10.1007/978-1-0716-4252-8_12,

MRI in assessing the progression of rare diseases like MCT8 deficiency becomes even more crucial as scarcity of samples is a common hindrance to research. It has been described that MCT8 deficiency courses with severe symptoms such as neurological impairments, delayed myelination, and psychomotor disorders, among others, with no effective treatment for the neurological component of the disease [13, 14]. That is why non-invasive in vivo techniques such as MRI are emerging as promising tools for their clinical and preclinical applications in studying neurological diseases. Its application to animal models is pivotal for both basic research and preclinical studies, particularly considering the current absence of effective therapeutic solutions for neurological impairments in MCT8 deficiency. In this context, a feasible and reliable non-invasive imaging modality offering tractable biomarkers of the disease greatly facilitates the obtention of mechanistic insights underlying the disease progression and drug discovery efforts in preclinical phases, while also reducing the required number of animals.

Among the plethora of potential data derived from MRI imaging, this chapter will serve as a close-up of two particular MRI applications of great interest, given that MCT8 alterations lead to both vasculature alterations [11] and myelin impairments [12]: time-of-flight magnetic resonance angiography (TOF-MRA) and diffusion tensor imaging (DTI).

TOF-MRA is a non-invasive technique developed for studying arterial and venous system integrity. It lacks ionizing radiation exposure and does not necessarily rely on contrast injections, thereby decreasing the risk of allergies. Furthermore, it possesses the capability to generate high-resolution volumetric images [15].

TOF-MRA operates by exploiting the inflow of unsaturated blood into a specific region of interest (ROI). The technique uses radiofrequency (RF) pulses and magnetic gradients to create contrast between the flowing blood and the surrounding static tissues. As a simplified breakdown of how TOF-MRA operates, firstly, a RF pulse is applied to the ROI. This pulse flips the protons of the hydrogen nuclei in the blood out of their natural alignment using a magnetic field, a process known as excitation. Then, magnetic gradients are applied to generate a magnetic echo sequence, which is used to measure the returning signal from the excited protons. These signals are processed to create an image. As unsaturated blood flows within the delimited area during each TR (repetition time), it creates a stronger signal compared to the one coming from the surrounding static tissues due to the consistent supply of unsaturated spins of the protons in the flowing blood, a process known as blood flow enhancement. Finally, a background suppression or denoising step is carried out to enhance the blood vessel signal and make it easier to discriminate from the surrounding tissues.

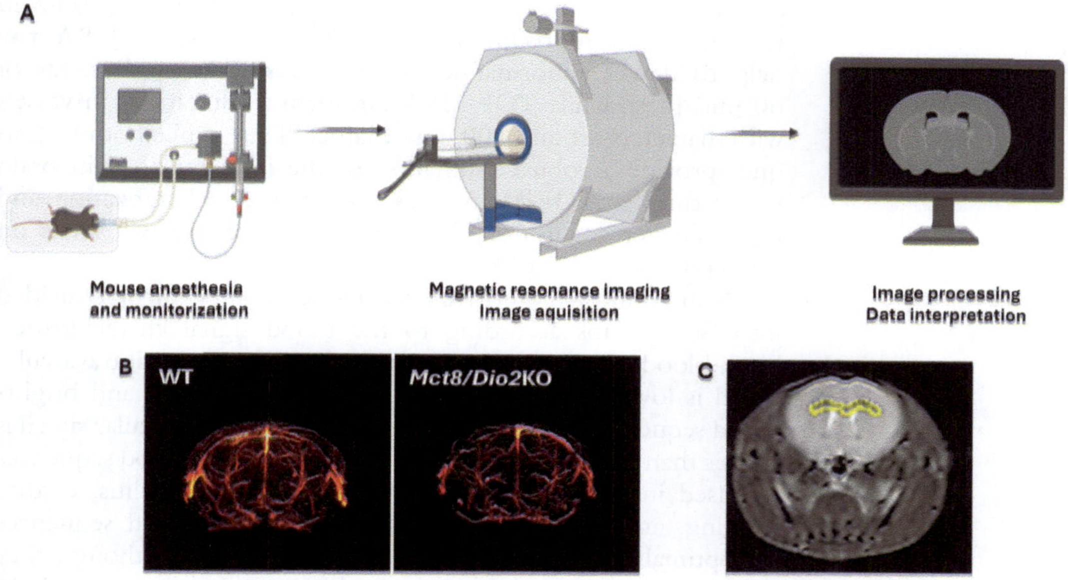

Fig. 1 Experimental procedures to explore MCT8 deficiency pathophysiology in vivo by magnetic resonance imaging techniques. (**a**) Representative protocol for magnetic resonance imaging procedures in mice. (**b**) The panel shows representative Pseudocolor Image Look-Up Table (LUT) maps from maximal intensity projections of 2D Time of Flight-Magnetic Resonance Angiography (TOF-MRA) acquisitions of the cerebral blood-vessel network of WT and *Mct8/Dio2*KO mice, an animal model for MCT8 deficiency. Modified from Guillén-Yunta M et al. Fluids and Barriers of the CNS (2023) 20:79. (**c**) T2 weighted image from a 6-month-old WT animal with embedded ROI for Diffusion Tensor Imaging (DTI) maps analyses

After all TR are acquired, the obtained TOF-MRA images undergo a maximum intensity projection (MIP) algorithm, in which pixels, representing blood flow, are extracted from the data set to recreate a 2D or 3D reconstruction of the blood vessels in a very short span of time (Fig. 1a, b). For cerebrovascular studies in particular, data acquisition with TOF-MRA may require 5 to 20 min to produce a high-resolution image in humans [16], while acquiring a whole-head reconstruction of the mouse brain vasculature in preclinical studies takes 30 to 40 min [11].

Thus, TOF-MRA and MIP images are often used to assess steno-occlusive lesions, arteriopathy, or cerebral hyperperfusion syndrome, making them very important clinical resources in the pediatric setting, as an efficient scan time is desirable to reduce sedation, particularly in young children who are prone to motion [17, 18].

TOF-MRA can also be acquired with a contrast agent injected into the vascular system to increase resolution (contrast-enhanced MRA or CE-MRA). Although CE-MRA is also employed for similar purposes as TOF-MRA, the advantages it offers over non-contrast methods are not always clear [16]. Furthermore, the acquisition time for CE-MRA is longer than TOF-MRA and requires subject catheterization, which can be triggering both in the clinics and in preclinical studies. Nevertheless, CE-MRA may help to detect abnormalities in small vessels (i.e., diameter of 60 μm) [19], while TOF-MRA can identify alterations in vessels with diameters around 100 μm or larger. Thus, while both techniques provide a robust estimation of the intracranial main brain vasculature, as TOF-MRA does not require catheterization and has a shorter acquisition time, in this chapter we will focus on non-contrast TOF-MRA.

Non-contrast TOF-MRA techniques can be further divided into two groups according to the blood signal characteristics: black-blood sequences (or dark-blood), where the intravascular signal is lower than that of the surrounding tissues; and bright-blood sequences (or white-blood), where the intravascular signal is higher than that of the surrounding tissues. Black-blood sequences are used in specific clinical scenarios, such as vasculitis, cardiac imaging, and susceptibility venography. Bright-blood sequences are optimal for studying the vascular lumen, and, although they have less accuracy in evaluating vessel walls in comparison to dark-blood techniques, they amplify the transverse magnetization of the protons in the blood using additional gradients, resulting in sequences with a high intravascular signal [20].

Regarding DTI, it has become one of the most popular techniques in modern brain research and clinical diagnostics. Since its inception in 1994, DTI has become a vital tool, spawning over 700 publications over the past decade alone. This technique enables the visualization and characterization of white matter fascicles, providing invaluable insights into normal brain function and various pathologies such as multiple sclerosis, stroke, aging, dementia, and schizophrenia [21].

DTI's principle lies in its ability to measure the displacement of water molecules within tissues, particularly highlighting the anisotropic nature of water diffusion in white matter, as water movement is fast along the fibers and slow perpendicular to them. By capturing the differential motion of water molecules along and perpendicular to nerve fibers, DTI extracts crucial diffusivity indices (Fig. 1a, b) [21].

Research suggests that myelin, the sheath enveloping axons, contributes significantly to DTI signals. From available DTI information, three parameters: fractional anisotropy (FA), axial diffusivity (AD), and radial diffusivity (RD) emerge as particularly reliable metrics when studying myelin status [22].

FA is a parameter that reflects the magnitude and orientation of anisotropic water diffusion from 0 to 1, predominantly reflecting the integrity of white matter fibers, with values nearing the unit in areas where white matter fibers are mostly ordered in the same sense [21]. This relation allows the detection of pathological events, which usually correspond with a decrease in this parameter

[22]. As for AD and RD, both parameters are related to the aforementioned differences in water movement in parallel or perpendicular directions to the ordered fibers. In physiological conditions, water diffusivity parallel to the fibers (AD) is greater than the diffusivity perpendicular to the fibers (RD). For this reason, demyelination results in an increased RD, presumably due to the lack or at least decrease of the barrier conformed by the myelin sheath, which typically hinders water diffusion perpendicular to axonal tracts. On the other hand, AD levels are much more connected to axonal integrity, thus axonal injury leads to decreased AD levels [23].

However, each of these parameters alone presents different individual hindrances, which can increase with anatomical and pathological complexity [21, 23]. For this reason, using only one of these parameters may not be enough to characterize tissue changes, and the integration of different measures yields a better interpretation of the underlying physiological status of myelin.

Despite its limitations and ongoing debates surrounding signal origins, DTI remains a cornerstone in understanding brain microstructure, particularly in the context of animal experimentation, where it allows non-invasive deep characterization of myelin status [12].

As research advances and methodologies evolve, MRI brain imaging continues to prove itself as a reliable way to evaluate the outcome of different pathophysiological processes. In particular, in MCT8 deficiency preclinical research, it allows for better monitoring of the different stages of the disease.

2 Materials

2.1 Time-of-Flight Magnetic Resonance Angiography (TOF-MRA)

1. A total number of animals between 8 and 10 individuals for each experimental group.
2. An inhalation anesthesia equipment filled with isoflurane.
3. Biospec spectrometer 7 T (Bruker Biospec 70/16 US (AV3); model number 1P BAP 70/16; serial number: S10051840, Bruker Biospin MRI GmbH, Ettlingen, Germany).
4. An MRI-compatible small animal gating system by SA Instruments (Stony Brook, NY).
5. A computer running Insight ToolKit (ITK-SNAP) free-licensed segmentation software (*see* **Note 1**).
6. A computer running ImageJ free-licensed image visualization software.
7. The Angio-MIP custom-made Image J plugin (*see* **Note 2**) or alternatively the 3D Volume Viewer tool from the ImageJ software.

8. A computer running Microsoft Excel.
9. A computer running statistics and graphing software such as GraphPad Prism.

2.2 Diffusion Tensor Imaging (DTI)

1. A total number of animals between 8 and 10 individuals for each experimental group.
2. An inhalation anesthesia equipment filled with isoflurane.
3. Biospec spectrometer 7 T (Bruker Biospec 70/16 US (AV3); model number 1P BAP 70/16; serial number: S10051840, Bruker Biospin MRI GmbH, Ettlingen, Germany), connected to a computer running Paravision 5.1 software (Bruker Medical GmbH).
4. An MRI-compatible small animal gating system by SA Instruments (Stony Brook, NY).
5. A computer running MATLAB (Mathworks, Natick, MA) software.
6. A computer running ImageJ free-licensed image visualization software.
7. A computer running Microsoft Excel.
8. A computer running statistics and graphing software such as GraphPad Prism.

3 Methods

3.1 TOF-MRA

1. Anesthesia and Monitoring:
   (a) Animals should be anesthetized with isoflurane via inhalation, and their temperature should be maintained at approximately 37 °C.
   (b) The physiological state of the animals should be monitored using an MRI-compatible small animal gating system, as MRI acquisitions require each animal to be anesthetized for a fair amount of time.
2. Image Acquisition: anesthetized animals are placed in the spectrometer for the acquisition of a 2D TOF-MRA without a contrast agent. Specific acquisition parameters are set: TE (echo time): 2.6 ms, TR: 15 ms, FA (flip angle): 80°, 80 slices of 0.3 mm without gap, matrix size 256 × 256, and FOV (field of view) 23 × 23 mm.
3. Image denoising: the acquired set of 80 whole-head TOF-MRA images undergoes a denoising process to enhance image quality and reduce artifacts (*see* **Note 3**).
4. Image processing: the denoised images are saved as a stack in the format .NIfTI using the ImageJ software. This format is

necessary for further segmentation analysis using the ITK-SNAP software (*see* **Note 4**). This step is unnecessary if you do not want to perform segmentation in your M.

5. RA images.
6. Image segmentation:
   (a) Open the stacked images with the ITK-SNAP software to perform segmentation of the specific whole-brain area in each of the 80 slices obtained per animal.
   (b) Adjust the brightness/contrast settings of the images to be able to clearly identify and segment the cerebral area.
   (c) The segmentation should be done manually with the Paintbrush Mode tool within ITK-SNAP with a recommended brush size = 3 (*see* **Note 5**).
   (d) Save the Segmentation Label in .NIfTI format.
   (e) If you want to segment different regions within the same stack of images, add another segmentation label and save it in .NIfTI format the same way.
   (f) Open ImageJ and upload the TOF-MRA stack and the Segmentation Label stack and then apply the segmentation to your images by multiplying the corresponding . NIfTI archives following these steps: Process>Image Calculator>Stack NIfTI · Multiply · Stack Label>Proceed. Save the resulting stack in .NifTI format.

   This step is not necessary if you don't want to do a segmentation in your TOF-MRA images.
7. Image analyses:
   (a) Run the Angio-MIP 2D plugin following these steps: Plugin>Analyze>Angio MIP 2D, applying the specific FOV conditions in the X, Y, and Z axis of your acquisition (i.e., FOV X = 23, Y = 23, Z = 24). The results will be provided as axial, coronal, and sagittal projections. Save the results in .NIfTI format.
   (b) In the obtained MIP projection, adjust the Threshold between 0 and 1 to delimit the specific brain-occupied region. This can be done following these steps: Image>Adjust>Threshold>Set Threshold>Set 0 · 1 > Apply.
   (c) With the wand tool, select the brain region and save the corresponding ROI with the ROI manager tool. This represents the whole brain-specific area.
   (d) Open the MIP projection image and establish a Threshold that allows you to specifically discriminate noise from vessel signal. Then, select the vessel-specific area following these steps: Edit>Selection>Create Selection (*see* **Note 6**)

and add this selection to the ROI Manager tool. This represents the cerebral blood vessels' specific area.

(e) Open the ROI Manager tool and overlay the previously created ROIs on the MIP projection image. Analyze the image by selecting Analyze>Measure (also CTRL+M). Save the results in an Excel document.

(f) Apply this formula:

$$\text{Cerebral blood vessel} - \text{covered area} = \frac{\text{blood vessel} - \text{specific area}}{\text{Segmented whole brain area}}$$

8. Statistical analyses: carry out the corresponding statistical analysis and graph your results using software such as GraphPad Prism.
9. Additionally, you can transform the black and white MIP projection images into color gradients or heatmaps by applying a Pseudocolor Image Look-Up Table (LUT) ImageJ plugin.

3.2 DTI

1. Anesthesia and Monitoring:
   (a) Animals should be anesthetized with isoflurane via inhalation, and their temperature should be maintained at approximately 37 °C.
   (b) The physiological state of the animals should be monitored using an MRI-compatible small animal gating system, as MRI acquisitions require each animal to be anesthetized for a fair amount of time.
2. Image Acquisition: anesthetized animals are placed in the spectrometer for the acquisition of two types of images:
   (a) T2-weighted (T2-W) spin-echo images were acquired using a rapid acquisition and relaxation enhancement (RARE) sequence in axial orientations, after setting the following parameters: TR = 2500 ms, TE = 44.3 ms, RARE factor = 8, Av = 3, FOV = 2.3 cm, acquisition matrix = 256 × 256 corresponding to an in-plane resolution of 90 × 90 m^2, slice thickness = 1.00 mm without gap and number of slices = 14.
   (b) DTI data were acquired using a spin echo single-shot echo planar imaging (EPI) pulse sequence with the following parameters: TR/TE 3500/40 ms; a signal average of 4.7, noncollinear diffusion gradient scheme with a diffusion weighting b = 110 and 1400 s/mm^2, slices thickness 1 mm without gap, FOV 23 × 23 mm and acquisition matrix = 128 × 128.
3. Image analyses:

(a) The different maps and numerical parameters derived from DTI acquisitions (FA, mean diffusivity, AD, RD trace, the eigenvalues, and eigenvector maps, etc.) can be calculated using software from MathWorks, Natick, MA (*see* **Note 7**).

(b) After this, a folder for each animal is generated containing its T2-weighted images and the different DTI maps corresponding to the whole brain for each animal.

(c) The first step before proceeding with the analysis is to identify the bregma coordinates of interest, corresponding to the particular brain area that will be studied. For this purpose, it is better to use the T2 weighted images, as they provide higher resolution compared to the DTI maps. To open T2-weighted files, simply drag and drop them into the ImageJ interface.

(d) Once the slices/images have been selected, the next step is to choose the ROI for each image. Different drawing tools can be used to outline the area of interest in each image. Then, save the ROI following these steps: Analyze>Tools>ROI Manager>Add (*see* **Note 8**).

(e) These ROIs will then be used to analyze the different DTI maps corresponding to each selected T2-weighted image. The raw DTI maps cannot be directly opened by dragging and dropping them into the ImageJ interface. Instead, they have to be imported using the following steps: File>Import>Raw>Select the map of interest>Adjust the parameters to the image (i.e., Image type = 64-bit Real, Width = 128, Height = 128, Offset to first image = 0 bytes, N of images = 1, Gap between images = 0, Tick little endian byte order) (*see* **Note 9**).

(f) After opening the images, overlay the previously created ROIs onto their corresponding images, by selecting a particular ROI. Then, analyze it by navigating to Analyze>Measure (or using the shortcut CTRL+M).

(g) From the obtained values, the mean gray value will provide information on water molecules' movement for each parameter (i.e., FA, AD, and RD).

4. Save the results in an Excel document.
5. Statistical analyses: carry out the appropriate statistical analysis and graph the results using software such as GraphPad Prism.

4 Notes

1. The segmentation step is necessary to avoid taking into account the peripheral blood vessels of the head-neck articulation, which are also captured in the whole-head MRA acquisition. This step will also allow you to delimit specific cerebral regions and analyze the blood vessel density in those areas, a task that cannot be done without a segmentation step.
2. The MIP2D plugin used in our study was made by specialists from the Magnetic Imaging Resonance Core Facility, Institut d'Investigacions Biomèdiques August Pi I Sunyer (IDIBAPS), Barcelona, Spain and the Laboratory of Surgical and Experimental Neuroanatomy, Faculty of Medicine and Health Sciences, Institute of Neurosciences, University of Barcelona, Barcelona, Spain.
3. The denoising step is necessary before image processing to diminish potential artifacts caused during the image acquisition, such as animal breathing or spontaneous reflexes. In our studies, the denoising process was conducted by specialists from the Magnetic Imaging Resonance Core Facility at the IDIBAPS, Barcelona, Spain.
4. ITK-SNAP is a software sensitive to character encoding; therefore, all folders related to MRA images should not contain special characters such as accents or the letter *ñ*.
5. This brush size allows the researcher to clearly define the cerebral area in each slice and avoid including a meningeal signal, which can introduce significant noise in the final measurements.
6. At this step, only the white-colored area corresponding to the cerebral blood vessels should be selected. If the black background of the image is automatically selected instead, go to Edit>Selection>Make inverse.
7. All the mentioned parameters were calculated with a custom software application developed in MATLAB at the Biomedical NMR Facility Sebastián Cerdán of the IIBm (CSIC-UAM).
8. TI maps include big numerical differences between white matter and grey matter values. These big differences make it imperative to accurately draw your ROI. In the event of not sufficient resolution on T2 images, it is preferable to ensure that the area enclosed by the ROI drawing is white matter, even if it means measuring a smaller area.
9. Image parameters may vary depending on the acquisition parameters, and using incorrect image parameters in ImageJ may yield inaccurate or wrong results. In case of uncertainty about

the acquisition parameters, it is advisable to consult the MRI expert responsible for the acquisitions.

References

1. Boxer AL, Rankin KP, Miller BL et al (2003) Cinguloparietal atrophy distinguishes Alzheimer disease from semantic dementia. Arch Neurol 60(7):949–956. https://doi.org/10.1001/archneur.60.7.949
2. Chetelat G, Baron J-C (2003) Early diagnosis of Alzheimer's disease: contribution of structural neuroimaging. NeuroImage 18(2): 525–541. https://doi.org/10.1016/s1053-8119(02)00026-5
3. Chen X, Roberts N, Zheng Q et al (2024) Comparison of diffusion tensor imaging (DTI) tissue characterization parameters in white matter tracts of patients with multiple sclerosis (MS) and neuromyelitis optica spectrum disorder (NMOSD), vol 34. Published online January, Eur Radiol, p 5263. https://doi.org/10.1007/s00330-023-10550-1
4. Krzyżak AT, Lasek J, Schneider Z et al (2024) Diffusion tensor imaging metrics as natural markers of multiple sclerosis-induced brain disorders with a low expanded disability status scale score. NeuroImage 290:120567. https://doi.org/10.1016/j.neuroimage.2024.120567
5. Matheus MG, Lehman RK, Bonilha L, Holden KR (2015) Redefining the pediatric phenotype of X-linked monocarboxylate transporter 8 (MCT8) deficiency: implications for diagnosis and therapies. J Child Neurol 30(12): 1664–1668. https://doi.org/10.1177/0883073815578524
6. Sijens PE, Rödiger LA, Meiners LC, Lunsing RJ (2008) 1H magnetic resonance spectroscopy in monocarboxylate transporter 8 gene deficiency. J Clin Endocrinol Metab 93(5): 1854–1859. https://doi.org/10.1210/jc.2007-2441
7. Kobayashi S, Onuma A, Inui T et al (2014) Clinical course and images of four familial cases of Allan-Herndon-Dudley syndrome with a novel monocarboxylate transporter 8 gene mutation. Pediatr Neurol 51(3): 414–416. https://doi.org/10.1016/j.pediatrneurol.2014.05.004
8. Thiessen JD, Glazner KAC, Nafez S et al (2010) Histochemical visualization and diffusion MRI at 7 tesla in the TgCRND8 transgenic model of Alzheimer's disease. Brain Struct Funct 215(1):29–36. https://doi.org/10.1007/s00429-010-0271-z
9. Genovese G, Palombo M, Santin MD et al (2021) Inflammation-driven glial alterations in the cuprizone mouse model probed with diffusion-weighted magnetic resonance spectroscopy at 11.7 T. NMR Biomed 34(4): e4480. https://doi.org/10.1002/nbm.4480
10. Petiet A, Aigrot M-S, Stankoff B (2016) Gray and white matter demyelination and Remyelination detected with multimodal quantitative MRI analysis at 11.7T in a chronic mouse model of multiple sclerosis. Front Neurosci 10:491. https://doi.org/10.3389/fnins.2016.00491
11. Guillén-Yunta M, Valcárcel-Hernández V, García-Aldea Á et al (2023) Neurovascular unit disruption and blood–brain barrier leakage in MCT8 deficiency. Fluids Barriers CNS 20(1): 79. https://doi.org/10.1186/s12987-023-00481-w
12. Valcárcel-Hernández V, López-Espíndola D, Guillén-Yunta M et al (2022) Deficient thyroid hormone transport to the brain leads to impairments in axonal caliber and oligodendroglial development. Neurobiol Dis 162:105567. https://doi.org/10.1016/j.nbd.2021.105567
13. Grijota-Martínez C, Bárez-López S, Gómez-Andrés D et al (2020) MCT8 deficiency: the road to therapies for a rare disease. Front Neurosci 14(April):1–8. https://doi.org/10.3389/fnins.2020.00380
14. Sarret C, Oliver Petit I, Tonduti D Allan-Herndon-Dudley syndrome. In: Adam MP, Ardinger HH, Pagon RA et al (eds) GeneReviews® Seattle Univ Washington, Seattle. Published online 2020
15. Yano T, Kodama T, Suzuki Y, Watanabe K (1997) Gadolinium-enhanced 3D time-of-flight MR angiography Experimental and clinical evaluation. Acta Radiol 38(1):47–54. https://doi.org/10.1080/02841859709171241
16. Townsend TC, Saloner D, Pan XM, Rapp JH (2003) Contrast material-enhanced MRA overestimates severity of carotid stenosis, compared with 3D time-of-flight MRA. J Vasc Surg 38(1):36–40. https://doi.org/10.1016/s0741-5214(03)00332-x
17. Sato K, Yamada M, Kuroda H et al (2016) Time-of-flight MR angiography for detection of cerebral hyperperfusion syndrome after

superficial temporal artery–middle cerebral artery anastomosis in moyamoya disease. Am J Neuroradiol 37(7):1244–1248. https://doi.org/10.3174/ajnr.A4715

18. Lanzman BA, Huang Y, Lee EH et al (2021) Simultaneous time of flight-MRA and T2* imaging for cerebrovascular MRI. Neuroradiology 63(2):243–251. https://doi.org/10.1007/s00234-020-02499-5

19. Klohs J, Baltes C, Princz-Kranz F et al (2012) Contrast-enhanced magnetic resonance microangiography reveals remodeling of the cerebral microvasculature in transgenic ArcAβ mice. J Neurosci 32(5):1705–1713. https://doi.org/10.1523/JNEUROSCI.5626-11.2012

20. De Leucio A, De Jesus O (2024) MR Angiogram. In: StatPearls. Published online

21. Assaf Y, Pasternak O (2008) Diffusion tensor imaging (DTI)-based white matter mapping in brain research: a review. J Mol Neurosci 34(1): 51–61. https://doi.org/10.1007/s12031-007-0029-0

22. Yano R, Hata J, Abe Y et al (2018) Quantitative temporal changes in DTI values coupled with histological properties in cuprizone-induced demyelination and remyelination. Neurochem Int 119:151–158. https://doi.org/10.1016/j.neuint.2017.10.004

23. Sun P, Murphy RKJ, Gamble P, George A, Song SK, Ray WZ (2017) Diffusion assessment of cortical changes, induced by traumatic spinal cord injury. Brain Sci 7(2):1–13. https://doi.org/10.3390/brainsci7020021

Part V

Zebrafish Model

Chapter 13

Experimentally Manipulating the Thyroid Hormone Axis in Zebrafish

Melody Autumn, Jenny Zeng, Isabella Ranieri, and Sarah K. McMenamin

Abstract

Thyroid hormone (TH) is an endocrine factor with a diverse array of developmental, metamorphic, and metabolic functions conserved across vertebrates. Zebrafish (*Danio rerio*) are a tractable model for endocrinology research, and recent research efforts focus on the roles of TH in zebrafish morphogenesis, growth and behavior. Several powerful approaches have been developed in zebrafish to modulate the TH axis and peripheral sensitivity to the hormone. These approaches include gain- and loss-of-function mutations that target components of the TH signaling pathways, as well as pharmacological treatments to modulate TH synthesis and availability. Here, we review some of these approaches for generating hypo- and hyperthyroid physiology and phenotypes during post-embryonic zebrafish development. In particular, we focus on a transgenic method of producing hypothyroid fish via metronidazole-based thyroid ablation. This approach can straightforwardly generate large numbers of hypothyroid individuals along with euthyroid sibling controls, and we survey some of the research applications in which this system has been used.

Key words Zebrafish, Thyroid hormone, Nitroreductase ablation, Hypothyroidism, Hyperthyroidism, Metronidazole

1 Introduction

Thyroid hormone (TH) is a globally circulating endocrine factor crucial for the development and function of organ systems across diverse vertebrate species [1, 2], including fishes [3–5]. Endocrine systems in zebrafish (*Danio rerio*) have numerous similarities with mammalian systems and can allow efficient investigation into hormonal regulation of development, growth, behavior, and metabolism [6–8]. Disrupting selected components of the TH axis in zebrafish allows researchers to test the roles of different aspects of TH signaling on physiological, cellular, and developmental

The original version of the chapter has been revised. A correction to this chapter can be found at https://doi.org/10.1007/978-1-0716-4252-8_17

Constanza Contreras-Jurado (ed.), *Thyroid Hormones: Methods and Protocols*, Methods in Molecular Biology, vol. 2876, https://doi.org/10.1007/978-1-0716-4252-8_13,

Corrected Publication 2025

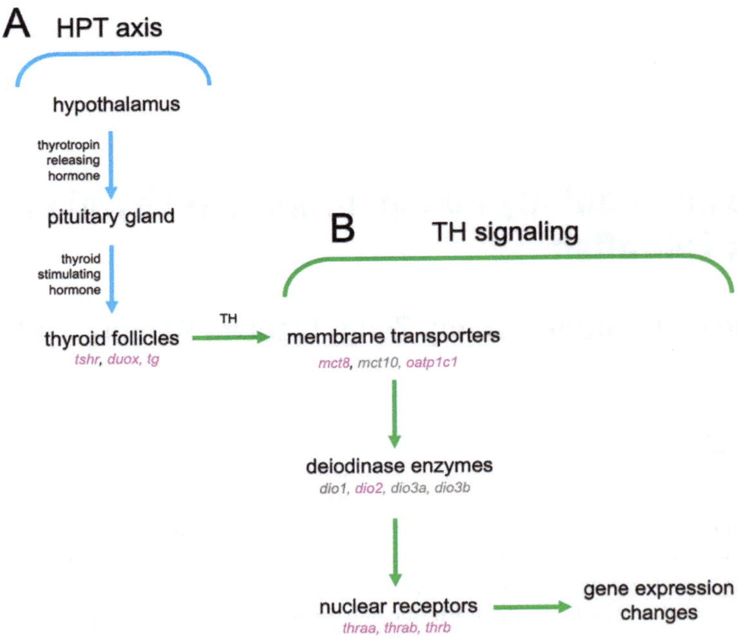

Fig. 1 Canonical thyroid signaling pathway. (**a**) The hypothalamus-pituitary-thyroid axis regulates synthesis of thyroid hormone. (**b**) Canonical TH signaling is mediated by nuclear hormone receptors that recruit histone-modifying enzymes to regulate gene expression. Selected genes involved in different steps of the pathways are indicated in italics; magenta color identifies genes for which mutant lines exist in zebrafish

processes. Here, we survey a few of the extensive genetic, transgenic, and pharmacological tools that allow specific disruption of specific components of TH signaling in zebrafish to produce functionally hypo- or hyperthyroid zebrafish during post-embryonic stages of development. Transgenic ablation of the thyroid follicles is a straightforward method for inducing hypothyroidism in zebrafish larvae [9], and we briefly present our lab's protocols for using this transgenic line.

TH circulates through the blood stream and must be actively transported into peripheral cells by membrane transporters, which regulate the amount of intracellular TH. Zebrafish have several TH transporters, each expressed in different tissues [10] (e.g., *mct8* in neurons [11]). In zebrafish, *mct8* [12], *mct10* [13], and *oatp1c1* [13] are the predominant membrane transporters [10]; loss-of-function mutant lines for *mct8* [14] and *oatp1c1* have been generated [15] (see Fig. 1 and Table 1).

TH is predominately produced in the form of T4; the bioactivity of this pro-hormone is modulated by deiodinase enzymes that remove iodine atoms from the molecule. Dio1 and Dio2 deiodinate T4 into the more genomically active form, T3 [16, 17]. A loss-of-function mutation in *dio2* functionally decreases the amount of

Table 1
Selected resources available in zebrafish to modulate thyroid hormone metabolism and pathway throughout post-embryonic development. Loss-of-function indicated as LOF; gain of function indicated as GOF

Category	Name	Type	Functional effect on TH metabolism	Citation (s)
TH transporter	*oatp1c1*	LOF mutant	Hypothyroidism	[15]
TH transporter	*mct8*	LOF mutant	Hypothyroidism	[15]
TH transporter	*slc16a2*	LOF mutant	Hypothyroidism	[14]
Deiodination	*dio2*	LOF mutant	Hypothyroidism	[18, 19]
TH receptor	*thraa*	LOF mutant	Hypothyroidism	[21, 22]
TH receptor	*thrab*	LOF mutant	Hypothyroidism	[21, 22]
TH receptor	*thrb*	LOF mutant	Hypothyroidism	[21]
Pituitary-thyroid axis	*tshb*	LOF mutant	Hypothyroidism	[14]
Pituitary-thyroid axis	*tshr*	LOF mutant	Hypothyroidism	[9]
Pituitary-thyroid axis	*tshr*	GOF mutant	Hyperthyroidism	[9]
TH synthesis	*thyroglobulin*	LOF mutant	Hypothyroidism	[14]
TH synthesis	*duox*	LOF mutant	Hypothyroidism	[29]
TH availability	T2	Drug	Hyperthyroidism	[49]
TH availability	T3	Drug	Hyperthyroidism	[50, 51]
TH availability	T4	Drug	Hyperthyroidism	[52]
TH availability	TU	Drug	Hypothyroidism	[23, 31]
TH availability	PTU	Drug	Hypothyroidism	[31, 51]
TH availability	MMI	Drug	Hypothyroidism	[31, 53]
TH availability, Deiodination	IOP	Drug	Hypothyroidism	[28]
TH availability	$KClO_4$	Drug	Hypothyroidism	[31]
TH availability, Deiodination	MPI cocktail	Drug	Hypothyroidism	[23, 33]
TH synthesis	Thyroid ablation	Transgenic	Hypothyroidism	[9]

bioactive T3, and two loss-of-function *dio2* mutant lines have been established [18, 19] (Table 1). The Dio3 enzyme deiodinates T4 into the largely inactive form rT3, and inactivating mutations in the zebrafish *dio3* isoforms *dio3a* and *dio3b* [20] functionally increase the bioavailability of the hormone.

Bioactive T3 preferentially interacts with nuclear TH receptors (THR). Zebrafish have several THR paralogues, and there are null mutations exist for several of the receptors [21, 22]. Work with these THR mutants has demonstrated that unliganded THRs have active roles in zebrafish development, in particular *thrab* null mutants can rescue certain tissue-specific effects of hypothyroidism, including aspects of pigmentation [21] and the proximodistal patterning of fin rays [23].

Mutants affecting downstream components of TH signaling obviously have variable effects on phenotypes in different organ systems and tissues, and these have been extremely informative in determining the functions of specific pathway components. Modulating global TH production causes the most severe and systemic phenotypes. The production of TH can be altered at the level of the thyroid follicles where the hormone is synthesized (fish do not have a discrete thyroid gland [24]).Thyroid-stimulating hormone (TSH) is produced by the pituitary and activates TSHR receptors (see Fig. 1). A *tshba* knockout mutant results in marked hypothyroid phenotypes [14], while a constitutively activating mutation in zebrafish *tshr* results in overproduction of TH and severe congenital hyperthyroidism [9].

Pharmacological treatments can directly alter thyroid hormone availability. Zebrafish can be straightforwardly treated with T2, T3 and/or T4 to activate TH signaling [25, 26]; T4 is typically the most stable and soluble. A concentration of 20–30 nM T4 is sufficient to induce hyperthyroid phenotypes in larvae [9]. Treatment with supplemental TH can also be used to rescue hypothyroid phenotypes [23, 27, 28].

TH is synthesized from the precursor thyroglobulin, expressed in the thyroid follicular cells [24]; the process requires the production of reactive oxygen species by the NADPH oxidase Duox. Loss of function mutations in *duox* prevent TH synthesis, resulting in hypothyroidism [29]. Several goitrogenic compounds can induce hypothyroidism by blocking the process of TH synthesis. These hypothyroidism-inducing compounds include thiourea (TU), its derivative 6-propyl-2-thiouracil (PTU) [30], and methimazole (MMI); these drugs all inhibit peroxidases needed for TH synthesis [31]. Potassium perchlorate ($KClO_4$) blocks iodine uptake in the thyroid follicles while iopanoic acid (IOP) blocks TH release and the T4-to-T3 conversion [32]. Finally, a cocktail (termed MPI) of multiple drugs can efficiently induce hypothyroidism in adult fishes [23, 33]; we have used this MPI cocktail in zebrafish at a concentration of 1 mM methimazole, 0.1 mM $KClO_4$, 0.01 mM IOP and 0.05% DMSO [23].

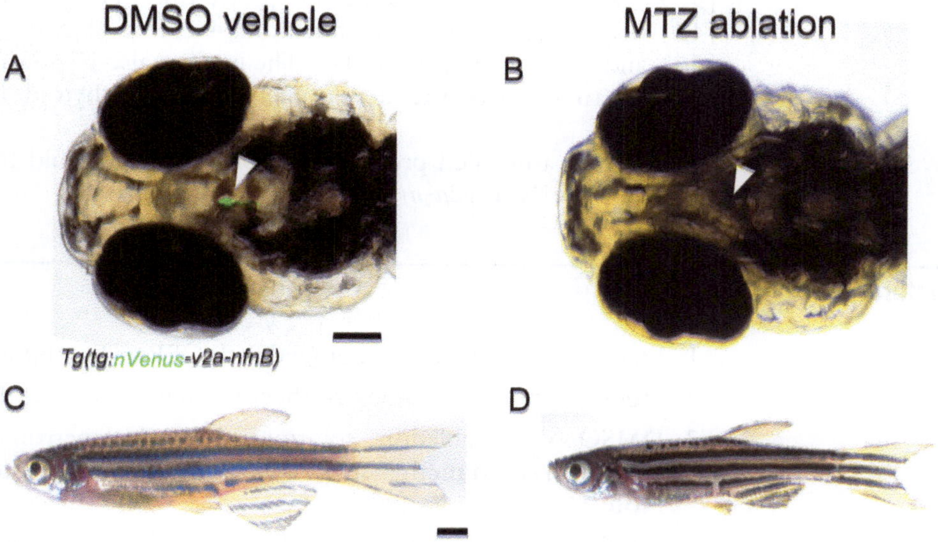

Fig. 2 Thyroid follicle ablation at 4 dpf. (**a–b**) Larvae at 5 dpf, 24 h after either (**a**) treatment with DMSO vehicle control or (**b**) ablation with MTZ. Note the loss of nVenus signal after thyroid follicles have been ablated. (**c–d**) Adult transgenic zebrafish that were treated with either (**c**) DMSO or (**d**) MTZ as larvae. Scale bars (**a–b**) 100 μm, (**c–d**) 2 mm

Chronic treatment with drugs requires frequent water changes, which can be disruptive to zebrafish. Transgenic ablation of the thyroid follicles prevents TH synthesis and results in robust hypothyroidism without the need for perpetual drug treatments (see Fig. 2). *Tg(tg:nVenus-v2a-nfnB)* is a metronidazole-inducible system where the thyroglobulin promotor [9] drives the expression of bacterial enzyme gene *nitroreductase* that converts the pro-drug metronidazole (MTZ) into cytotoxic byproducts [34]. This system permits the specific ablation of thyroglobulin-producing cells [9]. At larval stages, a single overnight treatment with MTZ eliminates the thyroid follicles, which do not regenerate [9]. This system can be used to produce large numbers of hypothyroid individuals as well as vehicle-treated sibling controls. Our lab performs the thyroid ablations at 3 and 4 days post fertilization (dpf), immediately after the thyroid follicles have developed with high efficiency. We note that thyroid ablation with this transgenic system is not typically effective after larval stages, and we have not been able to perform complete thyroid ablations in transgenic juveniles or adults.

Thyroid-ablated larvae grow into zebrafish with a number of informative phenotypic and physiological abnormalities (Fig. 2c–d). Hypothyroid zebrafish produced using the *Tg(tg:nVenus-v2a-nfnB)* line have been used to study developmental coordination [35], the development of the skeleton [23, 36–38], pigmentation

[9, 21, 39, 40], skin and scales [19], intestine [41], metabolism [42, 43], neural features and function [44, 45], photoreceptors [46], and feeding kinematics [47]. The line has also been used as a reporter to evaluate the developmental of the thyroid follicle itself [48].

Below is a detailed protocol for ablating the thyroid follicles using *Tg(tg:nVenus-v2a-nfnB)*.

2 Materials

1. Fluorescent microscope equipped with GFP or YFP filter cube.
2. Stock solution of tricaine or other anesthetic (*see* **Note 1**).
3. DMSO+MTZ ablation solution: 10 mM metronidazole (MTZ) in 1% dimethyl sulfoxide (DMSO) in fish water (*see* **Note 1**).
4. DMSO control solution: 1% DMSO in fish water. Store in light-protected bottle at 4C. (*see* **Note 1**).
5. Petri dishes, 22 mm deep.
6. Plastic transfer pipettes.
7. Fish system water.

3 Methods

1. Rear *Tg(tg:nVenus-v2a-nfnB)* larvae to either 3 or 4 dpf. Give larvae a single feed (of rotifers or powdered food) before beginning the treatment.
2. Warm aliquots of the DMSO and MTZ solutions up to 27C.
3. Anesthetize larvae with tricaine.
4. Using a low-magnification fluorescent microscope, sort larvae for nVenus-positive thyroid follicles (Fig. 1a) (*see* **Notes 2** and **3**).
5. If an MTZ-treated non-transgenic sibling control is desired, retain the nVenus-negative individuals for subsequent treatment with MTZ. (*see* **Note 4**).
6. With a transfer pipette, move the transgenic larvae into new dishes for MTZ and DMSO treatments (*see* **Note 5**).
7. Using a transfer pipette, remove as much liquid as possible from each dish and add pre-warmed (27C) MTZ ablation or DMSO mock-ablation solution.
8. Place dishes in a dark incubator (27C) and leave at least 8 h or overnight.

9. Remove as much solution as possible and replace with fish system water.
10. 24 h after initiating the ablation, screen larvae for nVenus expression. MTZ-treated larvae should show no residual nVenus expression (Fig. 1b).
11. Rear larvae on TH-free or low-TH foods (i.e., spirulina flakes; *Artemia*) to desired stage (*see* **Note 6**).

3.1 Troubleshooting

- If ablating at 4 dpf, it is advisable to feed live rotifers prior to the ablation, which can increase survival. We have found that larvae do not eat while in the MTZ solution, so the feeding should be performed prior to the treatment.
- If a few larvae retain Venus expression after the MTZ treatment, they can be either culled or else treated for a second overnight to fully ablate the thyroid tissues. The thyroid follicle tissues in these individuals may not have been fully developed before the initial ablation was performed.
- If most or all of the larvae retain Venus expression after the MTZ treatment, the MTZ solution may need to be remade or larvae can be left in the solution for longer amounts of time.

4 Notes

1. Store in light-protected bottles at 4 °C.
2. Transgenic insertions are maintained at a high copy number, and intercrosses of our transgenic stocks typically yield >75% positive F1s.
3. Although nVenus is yellow, it can be screened under most broad GFP filters.
4. Some slow-developing larvae may not have fully developed thyroid follicles 3 dpf. Take extra care in sorting and MTZ-ablating any Venus-negative siblings, because developmental delays could cause a transgenic larva to appear Venus-negative at 3 (or even 4) dpf, and these could be partially or fully ablated with the MTZ treatment.
5. Ensure that all treatments are kept at equivalent densities within dishes. Density should not exceed 20–25 larvae per 60 mL solution.
6. Commercially available fish food likely contains TH, which can lessen the severity of the hypothyroid phenotype.

Acknowledgments

We thank David Parichy in whose lab the transgenic thyroid ablation system was developed. We also thank all members of the McMenamin Lab and Boston College Animal Care Facility who contributed to fish husbandry, and to the anonymous reviewers for their comments.

Funding This work was supported by NSF CAREER 1845513 and NIH R35GM146467 (SM) as well as undergraduate support from the Boston College Vice Provost for Research.

References

1. Van Der Spek AH, Fliers E, Boelen A (2017) The classic pathways of thyroid hormone metabolism. Mol Cell Endocrinol 458:29–38. https://doi.org/10.1016/j.mce.2017.01.025
2. Zwahlen J, Gairin E, Vianello S, Mercader M, Roux N, Laudet V (2024) The ecological function of thyroid hormones. Philos Trans R Soc Lond B Biol Sci 379(1898):20220511. https://doi.org/10.1098/rstb.2022.0511
3. Campinho MA (2019) Teleost metamorphosis: the role of thyroid hormone. Front Endocrinol (Lausanne) 10:448915. https://doi.org/10.3389/FENDO.2019.00383/BIBTEX
4. McMenamin SK, Parichy DM (2013) Metamorphosis in teleosts. Curr Top Dev Biol 103:127–165. https://doi.org/10.1016/B978-0-12-385979-2.00005-8
5. Laudet V (2011) The origins and evolution of vertebrate metamorphosis. Curr Biol 21(18): R726–R737. https://doi.org/10.1016/J.CUB.2011.07.030
6. Löhr H, Hammerschmidt M (2011) Zebrafish in endocrine systems: recent advances and implications for human disease. Annu Rev Physiol 73:183–211. https://doi.org/10.1146/ANNUREV-PHYSIOL-012110-142320
7. Ghaddar B, Diotel N (2022) Zebrafish: a new promise to study the impact of metabolic disorders on the brain. Int J Mol Sc 23(10):5372. https://doi.org/10.3390/IJMS23105372
8. Porazzi P, Calebiro D, Benato F, Tiso N, Persani L (2009) Thyroid gland development and function in the zebrafish model. Mol Cell Endocrinol 312(1–2):14–23. https://doi.org/10.1016/J.MCE.2009.05.011
9. McMenamin SK et al (2014) Thyroid hormone-dependent adult pigment cell lineage and pattern in zebrafish. Science 345(6202): 1358–1361, Sep. 2014. https://doi.org/10.1126/science.1256251
10. A. M. Muzzio, P. D. Noyes, H. M. Stapleton, and S. C. Lema, "Tissue distribution and thyroid hormone effects on mRNA abundance for membrane transporters Mct8, Mct10, and organic anion-transporting polypeptides (Oatps) in a teleost fish," Comp Biochem Physiol A Mol Integr Physiol, vol. 167, pp. 77–89, Jan. 2014, doi: https://doi.org/10.1016/J.CBPA.2013.09.019
11. Zada D, Tovin A, Lerer-Goldshtein T, Appelbaum L (2016) Pharmacological treatment and BBB-targeted genetic therapy for MCT8-dependent hypomyelination in zebrafish. DMM Disease Models and Mechanisms 9(11):1339–1348. https://doi.org/10.1242/DMM.027227/257264/AM/PHARMACOLOGICAL-AND-BBB-TARGETED-GENETIC-THERAPIES
12. Arjona FJ, De Vrieze E, Visser TJ, Flik G, Klaren PHM (2011) Identification and functional characterization of zebrafish solute carrier Slc16a2 (Mct8) as a thyroid hormone membrane transporter. Endocrinology 152(12): 5065–5073. https://doi.org/10.1210/EN.2011-1166
13. Vatine GD et al (2012) Zebrafish as a model for Monocarboxyl transporter 8-deficiency. https://doi.org/10.1074/jbc.M112.413831
14. Song J et al (2021) Functions of the thyroid-stimulating hormone on key developmental features revealed in a series of zebrafish dyshormonogenesis models. Cells 10(8):1984. https://doi.org/10.3390/CELLS10081984/S1
15. I. Admati *et al.*, "Neural Alterations and Hyperactivity of the Hypothalamic–Pituitary–

Thyroid Axis in Oatp1c1 Deficiency," *https://home.liebertpub.com/thy*, vol. 30, no. 1, pp. 161–174, 2020, doi: https://doi.org/10.1089/THY.2019.0320

16. A. Orozco and C. Valverde-R, "Thyroid Hormone Deiodination in Fish," *https://home.liebertpub.com/thy*, vol. 15, no. 8, pp. 799–813, 2005, doi: https://doi.org/10.1089/THY.2005.15.799
17. Thisse C et al (2003) Spatial and temporal expression patterns of selenoprotein genes during embryogenesis in zebrafish. Gene Expr Patterns 3(4):525–532. https://doi.org/10.1016/S1567-133X(03)00054-1
18. Houbrechts AM, Delarue J, Gabriëls IJ, Sourbron J, Darras VM (2016) Permanent deiodinase type 2 deficiency strongly perturbs zebrafish development, growth, and fertility. Endocrinology 157(9):3668–3681. https://doi.org/10.1210/EN.2016-1077
19. Aman AJ, Kim M, Saunders LM, Parichy DM (2021) Thyroid hormone regulates abrupt skin morphogenesis during zebrafish postembryonic development. Dev Biol 477:205–218. https://doi.org/10.1016/J.YDBIO.2021.05.019
20. Heijlen M et al (2014) Knockdown of type 3 iodothyronine deiodinase severely perturbs both embryonic and early larval development in zebrafish. Endocrinology 155(4):1547–1559. https://doi.org/10.1210/EN.2013-1660
21. Saunders LM et al (2019) Thyroid hormone regulates distinct paths to maturation in pigment cell lineages. elife 8. https://doi.org/10.7554/ELIFE.45181
22. Han CR, Wang H, Hoffmann V, Zerfas P, Kruhlak M, Cheng SY (2021) Thyroid hormone receptor α mutations cause heart defects in zebrafish. Thyroid 31(2):315–326. https://doi.org/10.1089/THY.2020.0332
23. Harper M et al (2023) Thyroid hormone regulates proximodistal patterning in fin rays. Proc Natl Acad Sci 120(21):e2219770120. https://doi.org/10.1073/PNAS.2219770120
24. Alt B, Reibe S, Feitosa NM, Elsalini OA, Wendl T, Rohr KB (2006) Analysis of origin and growth of the thyroid gland in zebrafish. Dev Dyn 235(7):1872–1883. https://doi.org/10.1002/DVDY.20831
25. Walpita CN, Crawford AD, Janssens EDR, Van Der Geyten S, Darras VM (2009) Type 2 iodothyronine deiodinase is essential for thyroid hormone-dependent embryonic development and pigmentation in zebrafish. Endocrinology 150(1):530–539. https://doi.org/10.1210/EN.2008-0457
26. Walpita CN, Van der Geyten S, Rurangwa E, Darras VM (2007) The effect of 3,5,3′-triiodothyronine supplementation on zebrafish (Danio rerio) embryonic development and expression of iodothyronine deiodinases and thyroid hormone receptors. Gen Comp Endocrinol 152(2–3):206–214. https://doi.org/10.1016/J.YGCEN.2007.02.020
27. Sekimizu K, Tagawa M, Takeda H (2007) Defective fin regeneration in Medaka fish (Oryzias latipes) with hypothyroidism. Zool Sci 24(7):693–699. https://doi.org/10.2108/zsj.24.693
28. Farias-Serratos BM, Lazcano I, Villalobos P, Darras VM, Orozco A (2021) Thyroid hormone deficiency during zebrafish development impairs central nervous system myelination. PLoS One 16(8):e0256207. https://doi.org/10.1371/JOURNAL.PONE.0256207
29. K. Chopra, S. Ishibashi, and E. Amaya, "Zebrafish duox mutations provide a model for human congenital hypothyroidism," Biol Open, vol. 8, no. 2, 2019, doi: https://doi.org/10.1242/bio.037655
30. Van Der Ven LTM, Van Den Brandhof EJ, Vos JH, Power DM, Wester PW (2006) Effects of the antithyroid agent propylthiouracil in a partial life cycle assay with zebrafish. Environ Sci Technol 40(1):74–81. https://doi.org/10.1021/ES050972C/ASSET/IMAGES/LARGE/ES050972CF00004.JPEG
31. Brown DD (1997) The role of thyroid hormone in zebrafish and axolotl development. Proc Natl Acad Sci USA 94(24):13011–13016. https://doi.org/10.1073/PNAS.94.24.13011/ASSET/369D75FD-B743-485A-A751-CB8E87C7BA1F/ASSETS/GRAPHIC/PQ2473226005.JPEG
32. Horie Y, Sawada A, Dorcas U, Ramaswamy BR, Iguchi T (2024) Iopanoic acid alters thyroid hormone-related gene expression, thyroid hormone levels, swim bladder inflation, and swimming performance in Japanese medaka. Comp Biochem Physiol, Part C: Toxicol Pharmacol 281:109930. https://doi.org/10.1016/J.CBPC.2024.109930
33. Salis P et al (2021) Thyroid hormones regulate the formation and environmental plasticity of white bars in clownfishes. Proc Natl Acad Sci USA 118(23):e2101634118. https://doi.org/10.1073/PNAS.2101634118/SUPPL_FILE/PNAS.2101634118.SAPP.PDF
34. Curado S, Stainier DYR, Anderson RM (2008) Nitroreductase-mediated cell/tissue ablation in zebrafish: a spatially and temporally controlled ablation method with applications in developmental and regeneration studies. Nat

Protoc 3(6):948–954. https://doi.org/10.1038/nprot.2008.58

35. Hu Y, Mauri A, Donahue J, Singh R, Acosta B, McMenamin S (2019) Thyroid hormone coordinates developmental trajectories but does not underlie developmental truncation in danionins. Dev Dyn 248(11):1144–1154. https://doi.org/10.1002/DVDY.76
36. Keer S, Cohen K, May C, Hu Y, McMenamin S, Hernandez LP (2019) Anatomical assessment of the adult skeleton of zebrafish reared under different thyroid hormone profiles. Anat Rec 302(10):1754–1769. https://doi.org/10.1002/ar.24139
37. Keer S et al (2022) Thyroid hormone shapes craniofacial bones during postembryonic zebrafish development. Evol Dev 24(1–2):61–76. https://doi.org/10.1111/EDE.12399
38. Galindo D et al (2019) Thyroid hormone modulation during zebrafish development recapitulates evolved diversity in danionin jaw protrusion mechanics. Evol Dev 21(5): 231–246. https://doi.org/10.1111/EDE.12299
39. Lewis VM et al (2019) Fate plasticity and reprogramming in genetically distinct populations of danio leucophores. Proc Natl Acad Sci USA 116(24):11806–11811. https://doi.org/10.1073/PNAS.1901021116/SUPPL_FILE/PNAS.1901021116.SD02.XLS
40. Eom DS, Bain EJ, Patterson LB, Grout ME, Parichy DM (2015) Long-distance communication by specialized cellular projections during pigment pattern development and evolution. elife 4(December 2015). https://doi.org/10.7554/ELIFE.12401
41. Blitz E et al (2021) Thyroid hormones regulate goblet cell differentiation and Fgf19-Fgfr4 signaling. Endocrinology 162(5):1–13. https://doi.org/10.1210/ENDOCR/BQAB047
42. Matsuda H (2018) Zebrafish as a model for studying functional pancreatic β cells development and regeneration. Develop Growth Differ 60(6):393–399. https://doi.org/10.1111/dgd.12565
43. Matsuda H, Mullapudi ST, Zhang Y, Hesselson D, Stainier DYR (2017) Thyroid hormone coordinates pancreatic islet maturation during the zebrafish larval-to-juvenile transition to maintain glucose homeostasis. Diabetes 66(10):2623–2635. https://doi.org/10.2337/DB16-1476
44. Walter KM, Miller GW, Chen X, Harvey DJ, Puschner B, Lein PJ (2019) Changes in thyroid hormone activity disrupt photomotor behavior of larval zebrafish. Neurotoxicology 74:47–57. https://doi.org/10.1016/J.NEURO.2019.05.008
45. Wasserman-Bartov T, Admati I, Lebenthal-Loinger I, Sharabany J, Lerer-Goldshtein T, Appelbaum L (2022) Tsh induces Agrp1 neuron proliferation in Oatp1c1-deficient zebrafish. J Neurosci 42(44):8214–8224. https://doi.org/10.1523/JNEUROSCI.0002-22.2022
46. Mackin RD et al (2019) Endocrine regulation of multichromatic color vision. Proc Natl Acad Sci USA 116(34):16882–16891. https://doi.org/10.1073/PNAS.1904783116/SUPPL_FILE/PNAS.1904783116.SM01.MP4
47. McMenamin SK, Carter C, Cooper WJ (2017) Thyroid hormone stimulates the onset of adult feeding kinematics in zebrafish. Zebrafish 14(6):517. https://doi.org/10.1089/ZEB.2017.1453
48. Gillotay P et al (2020) Single-cell transcriptome analysis reveals thyrocyte diversity in the zebrafish thyroid gland. EMBO Rep 21(12): e50612. https://doi.org/10.15252/EMBR.202050612/SUPPL_FILE/EMBR202050612-SUP-0001-APPENDIX.PDF
49. Lazcano I, Rodríguez-Ortiz R, Villalobos P, Martínez-Torres A, Solís-Saínz JC, Orozco A (2019) Knock-down of specific thyroid hormone receptor isoforms impairs body plan development in zebrafish. Front Endocrinol (Lausanne) 10(MAR):441653. https://doi.org/10.3389/FENDO.2019.00156/BIBTEX
50. Shkil F, Siomava N, Voronezhskaya E, Diogo R (2019) Effects of hyperthyroidism in the development of the appendicular skeleton and muscles of zebrafish, with notes on evolutionary developmental pathology (Evo-Devo-path). Scientific Reports 9(1):1–13. https://doi.org/10.1038/s41598-019-41912-9
51. Walter KM, Miller GW, Chen X, Yaghoobi B, Puschner B, Lein PJ (2019) Effects of thyroid hormone disruption on the ontogenetic expression of thyroid hormone signaling genes in developing zebrafish (Danio rerio). Gen Comp Endocrinol 272:20–32. https://doi.org/10.1016/J.YGCEN.2018.11.007
52. Marelli F et al (2016) Patterns of thyroid hormone receptor expression in zebrafish and generation of a novel model of resistance to thyroid hormone action. Mol Cell Endocrinol 424:102–117. https://doi.org/10.1016/J.MCE.2016.01.020
53. Rodrigues MS et al (2022) Thyroid hormones deficiency impairs male germ cell development: a cross talk between hypothalamic-pituitary-thyroid, and-gonadal axes in zebrafish. Front Cell Dev Biol 10. https://doi.org/10.3389/FCELL.2022.865948

Part VI

Clinical Studies

Chapter 14

Review of Surgical Interventions in the Thyroid Gland: Recent Advances and Current Considerations

Norberto Herrera Merino and María Victoria Redondo Vega

Abstract

The thyroid gland, located at the base of the neck, regulates metabolism and hormone balance through hormones like T4 and T3, which are essential for growth, neurological development, and energy production. Thyroid diseases affect 10% of the global population, making accurate and up-to-date information on surgical interventions and advancements crucial for improving clinical outcomes. Thyroid gland surgery is a dynamic field that has experienced remarkable advances in diagnosis, surgical techniques, and postoperative management. These include new advances in surgical techniques that improve precision, reduce surgical trauma, and speed up patient recovery, identification of biomarkers, and understanding of the molecular characteristics of tumors that allow for more targeted therapeutic strategies, and incorporation of advanced technologies that improve diagnostic accuracy and efficacy. This review aims to guide healthcare professionals and lay the groundwork for future research and innovative treatments in thyroid surgery.

Key words Thyroid gland, Thyroid surgery, Surgical techniques, Thyroid diseases, Nodular diseases

1 Introduction

The thyroid gland is a small but crucial endocrine gland located at the base of the neck. It plays a key role in regulating metabolism and hormone balance by producing thyroid hormones such as thyroxine (T4) and triiodothyronine (T3). These hormones are essential for controlling growth, the development of the neurological system, heart rate, body temperature, energy production, growth, and development. An imbalance in the production of these hormones can have significant health consequences.

The importance of this review is driven by the high prevalence of thyroid diseases, affecting 10% of the world's population according to the World Health Organization, as well as the prevalence of thyroid cancer [1]. It's essential to have up-to-date and accurate information on surgical interventions, procedures, technical advances, and long-term outcomes to guide healthcare

Constanza Contreras-Jurado (ed.), *Thyroid Hormones: Methods and Protocols*, Methods in Molecular Biology, vol. 2876,
https://doi.org/10.1007/978-1-0716-4252-8_14,

professionals and improve clinical outcomes. This review also aims to provide a solid foundation for future research and development of innovative therapeutic approaches in the field of thyroid surgery.

2 Thyroid Diseases

Thyroid diseases can be classified into four main groups, although they may occur together:

1. Disorders that affect the normal functioning of the gland:

Hypothyroidism Occurs when the thyroid does not produce enough thyroid hormones. This can result in decreased metabolism, fatigue, weight gain, cold intolerance, and other progressive symptoms that can lead to death in extreme cases.

Hyperthyroidism Characterized by excessive production of thyroid hormones, which can speed up metabolism. Symptoms may include unintentional weight loss, nervousness, excessive sweating, and increased heart rate. Graves' disease, an autoimmune disease defined by hyperthyroidism secondary to inappropriate production of antibodies against TSH receptors [2, 3], is the most common cause of hyperthyroidism in the USA.

The prevalence of thyroid dysfunction diseases has increased significantly in recent decades, although its exact value is difficult to know due to a high prevalence of undiagnosed thyroid dysfunction (between 3.8 and 5.3%) [4, 5], a high percentage of which is subclinical (80%).

Thyroiditis Inflammation of the thyroid gland, which can be caused by a variety of reasons, such as infections or autoimmune disorders.

Nodular disease Abnormal growths in the thyroid gland in the form of a single (thyroid nodule) or multiple nodular formations (multinodular goiter). The prevalence is 10 times higher in females, and depends on the type of study of the population: 4–8% by palpation, 20–50% by ultrasound (depending on age), and 50% in necropsies [6–8]. Its importance lies in the incidence of thyroid cancer presenting as nodular disease, so that the prevalence of thyroid cancer in the unselected population ranges from 1 to 5% [9]. The evaluation is based on a complete medical history, family history, assessment of thyroid function, ultrasound of the neck (thyroid gland and central and lateral compartments) to classify the ultrasound risk of malignancy (EU-TIRADS 1 to 5), ultrasound-guided fine needle aspiration biopsy of the nodule whose indication is defined by clinical, laboratory and ultrasound factors (in EU-TIRADS 5: >10 mm; EU-TIRADS 4: >15 mm; EU-TIRADS 3: >20 m) which allows the application of the Bethesda classification as follows:

Bethesda I: Non-diagnostic.

Bethesda II: Benign.

Bethesda III: Atypia of uncertain significance or follicular lesion of uncertain significance.

Bethesda IV: Follicular neoplasia or suspected follicular neoplasia.

Bethesda V: Suspected malignancy.

Bethesda VI: Malignant.

This allows us to establish the indication for conservative treatment and follow-up or surgical treatment [10], which we will specify later.

Thyroid Tumours Although less common, with an incidence in the European registry three times higher in women, 9.3 and 3.1 cases per 100,000 people/year, respectively [11], it can come from any of the cells that form it:

Thyroid Cancer They can be grouped into two large blocks, differentiated CDT (papillary carcinoma, follicular carcinoma, Hürthle cell carcinoma, and poorly differentiated carcinoma), which account for 90% of thyroid cancers, or anaplastic, which have a very poor prognosis and account for 1–2% of thyroid tumors, originating from epithelial cells [12, 13].

Molecular analysis of differentiated tumors allows classification into BRAF-predominant and RAS-predominant mutations [14] although there is no single pattern of mutations. Anaplastic tumors are usually BRAF or RAS mutations associated with other genomic alterations (TERT promoters, mTOR pathway, among others), and are allowing the development of targeted therapies [15].

With an incidence of 1/10,000 in Spain of approximately 3000 cases/year [16].

Medullary Carcinoma These arise from the "C" cells and account for around 5% of thyroid tumors. About 25% are hereditary (mainly familial or as part of MEN 2 A and B syndrome, Cowden, familial polyposis) in which there is a RET mutation, and in sporadic cases, more than 90% have RET and RAS mutations [17, 18]. The RET mutation is associated with an increased likelihood of metastatic disease.

Other Non-endocrine Tumors The most frequent are squamous cell carcinoma and lymphoma, in a percentage of less than 1%.

Metastatic Tumors Infrequent, except in autopsy series [19], the most frequent origins are primary malignant neoplasms of the kidney, breast, and lung [20].

3 Types of Surgery on the Thyroid Gland

Surgical interventions on the thyroid gland vary according to the nature and extent of the condition. Some of the common types of surgical procedures are described here:

1. Total Thyroidectomy

   Total thyroidectomy is a procedure in which the entire thyroid gland is removed leaving no macroscopic residue. This intervention is performed in several situations, mainly in thyroid cancer and in hyperthyroidism associated with severe disease or medically untreatable drug-induced hyperthyroidism.

2. Lobectomy (Hemithyroidectomy)

Lobectomy One of the two lobes of the thyroid gland is removed without leaving residual gross tissue. This approach is used when the disease is localized to a single lobe or when dealing with specific thyroid nodules as it allows a definitive histological diagnosis and if it requires surgery to complete total thyroidectomy it is not necessary to re-intervene on that side which is technically much more complex and with a high complication rate.

3. Partial thyroidectomy:

   These procedures involve the removal of part of the thyroid gland. They are considered in situations where the condition is localized to a specific part of the thyroid gland and does not affect the whole organ and are intended to reduce the chances of recurrent nerve damage or the need for postoperative hormone replacement therapy:

Nodulectomy Very infrequent and practically in disuse, it is only indicated in hyperfunctioning nodules (toxic adenoma) of small size, no more than 2 cm, which cannot be treated with other forms of hyperthyroidism control and without suspicion of malignancy.

Lobectomy/Partial Thyroidectomy This is performed when it is necessary to eliminate a significant portion of the gland, but complete removal is not required. It is mainly indicated in surgery for autoimmune hyperthyroidism and in multinodular goiter in order to reduce the need for thyroid hormone in the postoperative period.

There are two types:

- Near total thyroidectomy: Thyroid tissue is removed leaving less than 1 gram in the region adjacent to the arrival of the recurrent nerve in the cricothyroid muscle [21].
- Subtotal thyroidectomy: Thyroid tissue is removed leaving between 6 and 10 grams on one or both sides [21].

4 Key Points of Thyroid Surgery

Care should always be taken in the surgical assessment of the patient, as in any intervention, and in thyroid interventions, the function of the vocal cords should also be assessed by laryngoscopy, in case of voice alteration due to the pathology and in re-interventions on the gland to rule out sequelae of previous surgery [22].

The general principles that should guide all thyroid surgery are the following [23]:

- Know the surgical anatomy of the thyroid and the anatomical variants of the perithyroid structures, for thyroid and superior and recurrent laryngeal nerves.
- Correctly expose the gland.
- Do not perform any aggressive technical gestures without prior identification of the main anatomical structures.

Main anatomical structures.

- Localization of the laryngeal recurrent nerve.
- Be demanding with hemostasis in order to achieve a bloodless field.
- Avoid hemostasis with power instruments in areas close to the laryngeal nerves.
- A standardized, delicate, and meticulous technique.

Neuromonitoring of the recurrent laryngeal nerve is increasingly used in thyroid surgery due to anatomical variations, although it is not a substitute for a delicate and thorough technique and is routinely recommended by multiple authors. It should always be performed in total thyroidectomy and reinterventions.

After placing the patient in the supine position with the thorax elevated 30 degrees and the neck hyperextended, an incision is made in a cervical fold between the two sternocleid muscles (in the case of large goiters it may be larger) 2–3 cm from the jugulum; this is Kocher's cervical incision and is the "gold standard" [24].

In cases of re-interventions, a lateral access can be made in front of the ECM (lateral approach) with section of the sternohyoid muscle to locate the laryngeal recurrent nerve in the tracheo-esophageal groove or lateral to it depending on the side [22].

5 Complications and Postoperative Considerations in Thyroid Gland Surgery

Thyroid gland surgery is generally safe, but as with any surgical procedure, there are associated risks and complications. It is important that patients are informed about these risks and that the medical team provides adequate postoperative follow-up. Some relevant aspects are detailed here [25, 26]:

1. Complications associated with Thyroid Gland Surgery:

Recurrent Laryngeal Nerve Damage Recurrent laryngeal nerves, responsible for controlling the vocal cords, can be damaged during surgery, which could result in temporary or permanent voice changes, with bilateral paralysis being a very serious complication that may require tracheotomy for respiratory distress. It is the second most common complication, accounting for 20% of the total, and may be transient or permanent if it persists for more than 6 months [27].

Parathyroid Gland Damage The parathyroid glands, which regulate calcium levels in the body, can be affected, which may result in hypoparathyroidism (low parathyroid function) and low blood calcium levels. It is the most common complication of thyroid surgery, 15–30% transient and 1–7% permanent [28].

Bleeding Although rare (0.1 to 1% of thyroidectomies), excessive bleeding during or after surgery is a possible complication, most frequent in the first 6 hours [29–31].

Infection Infection at the surgical site is a potential complication, although it is minimized with aseptic surgical techniques and the use of antibiotics.

Aesthetic Problems Scarring can affect aesthetic appearance, especially in procedures requiring larger incisions.

Hypothyroidism or Hyperthyroidism Depending on the extent of surgery, patients may experience insufficient thyroid function (hypothyroidism) or excessive thyroid function (hyperthyroidism) after surgery.

2. Postoperative Management and Follow-Up:

Hormone Replacement In cases of total or partial thyroidectomy, hormone replacement may be necessary to maintain proper hormone balance in the body.

Calcium Level Monitoring Regular monitoring of blood calcium levels is performed to detect any signs of hypoparathyroidism.

Vocal Physiotherapy In cases of voice change due to recurrent laryngeal nerve damage, vocal physiotherapy may be beneficial for recovery.

Scar Care Instructions are provided on proper scar care to minimize the appearance of the scar.

Oncological Follow-up In cases of thyroid cancer, regular oncological follow-up is performed to detect any signs of recurrence.

Thyroid Function Assessment Thyroid function is monitored to ensure proper hormone balance, adjusting medication as needed.

Consultation with an Endocrinologist In many cases, regular consultation with an endocrinologist is recommended for ongoing management of thyroid function.

6 Indications for Thyroid Surgery

The decision to perform surgery on the thyroid gland is based on a variety of indications, which may include the presence of thyroid nodules or goiter and thyroid cancer, among others. The specific indications and their associated characteristics for surgically treated diseases are discussed below:

1. Thyroid nodules

The indications for surgical treatment of thyroid nodules are, according to the 2023 European Thyroid Association Clinical Practice Guidelines for thyroid nodule management [10]:

- Symptomatic nodular disease.
- Nodules classified as benign on complete work-up that become symptomatic over time.
- Calcitonin levels above the normal range.
- Increased calcitonin after stimulation test in RET mutation carriers.
- Nodules with indeterminate cytology (Bethesda class III and IV), which cannot be followed up adequately.
- Nodules with Bethesda class V and VI cytology.

Type of Surgery
In benign thyroid nodules (Bethesda II) surgical lobectomy is indicated in symptomatic, usually larger than 4 cm, hyperfunctioning, or intrathoracic. In nodules of uncertain cytology (Bethesda III-V) surgery is indicated [32], in single nodules the minimum surgery is a hemithyroidectomy, in case of multinodular goiters a near total or subtotal thyroidectomy may be indicated [33, 34].

2. Thyroid Cancer and Risk Factors:

Thyroid cancer is primarily treated by surgery. Risk factors and reasons for surgical intervention include:

Malignant Biopsy Results When biopsy of a thyroid nodule confirms the presence of cancerous cells, surgery is a primary option to remove the cancerous tissue, with differentiated thyroid cancer accounting for 95% of cases [35]. Total thyroidectomy is the treatment of choice, with the addition of central and lateral lymphadenectomy if there is lymph node involvement.

Size and Extent Surgery may be recommended if the thyroid cancer is large or has spread beyond the thyroid gland, to prevent spread to other areas of the body.

Age and General Health of the Patient The decision to perform surgery may also depend on the patient's age and general health, considering the risks and benefits of the intervention.

3. Multinodular goiter:

It should always be performed after information on the risks and results of the different treatment modalities, with radioactive iodine, thyroid hormone or simply to rule out malignancy as in nodules and surveillance [36]. The clear indications for surgery in multinodular goiter are:

- Compression: patients presenting with features of tracheal or esophageal compression or venous flow obstruction.
- Suspected malignancy.
- Cosmetic problems with patient preference.
- Symptomatic intrathoracic goiter.

4. Hyperthyroidism

Surgery is indicated in two possible situations [37, 38]:

- Toxic adenoma: in this case the minimum surgery is a lobectomy.
- Graves' disease: This should always be performed after information on the risks and results of the different treatment modalities as the results are similar. Surgery with total or subtotal thyroidectomy is indicated in patients with compressive or retrosternal symptoms, severe ophthalmopathy, suspicion of malignancy, patients who do not want to be treated with radioactive iodine or pregnant women who need rapid control of thyroid function.

7 Techniques and Recent Advances in Thyroid Surgery

1. Technological and pharmaceutical advances leading to safer surgery.

The three main complications of thyroid surgery are bleeding, intraoperative or postoperative, hypocalcemia and laryngeal nerve damage, and recent advances in thyroid surgery aim to reduce their incidence [39]:

- Bleeding:

  Topical hemostatic agents: in two formulations, gels or gauze type. These can be active by binding to soft tissues by covalent bonds or by carrying procoagulant agents (fibrinogen, thrombin, etc.) [40–42]. Their use does not significantly reduce postoperative bleeding, but they reduce drainage volume after surgery [43].

  Energy-based hemostasis instruments: with ultrasound (Harmonic Scalpel), bipolar or combined energy. They reduce surgery time and intraoperative bleeding [44].

- Postoperative hypocalcemia:

Localization of the glands and their preservation is essential to avoid injuring them.

- Fluorescence for localization and assessment of the vascularization of the parathyroid glands in thyroidectomy [45]. This can be done by introducing indocyanine green or by autofluorescence [46]. Its use has been shown to reduce postoperative hypocalcemia, but not permanent hypoparathyroidism and at the cost of increased operative time [28].
- Subcapsular injection of saline prior to parathyroid dissection. Particularly of the upper parathyroids (SCASI SubCApsular Saline Injection), described by Choi in 2018 [47], allows a reduction in transient hypoparathyroidism.
- Laryngeal nerve injury:

Visual localization using a standardized technique at anatomical landmarks along their pathway is the Gold standard [48].

- Electromyographic monitoring of the recurrent laryngeal nerves, now routinely indicated [49].

  Intermittent, at the surgeon's request, based on amplitude and latency of intermittent stimulation electromyographic response.

  Continuous vagus nerve monitoring to detect injurious maneuvers of the recurrent nerves [50].

Its efficacy is greater in less experienced surgeons than in experienced surgeons [51] and the introduction of protocols for its use in total thyroidectomy by stopping surgery in case of loss of signal after the first lobe is performed reduces the incidence of bilateral recurrent injury [50, 52].

It should always be borne in mind that all these advances only complement a refined and careful surgical technique with vascularization and dissection of the peri-thyroid structures, identification of the laryngeal nerves and parathyroid glands, avoiding injury to these by heat from hemostasis instruments.

2. New surgical techniques

Surgery of the thyroid gland has undergone significant advances in recent decades, with the development of more precise and less invasive techniques. Some of the recent techniques and advances in this field are detailed here:

Minimally Invasive, Endoscopy-Assisted Surgery

Minimally invasive, endoscopy-assisted surgery (MIVAT (minimally invasive video-assisted thyroidectomy)), which involves the use of small incisions and an endoscopic camera to visualize and access the thyroid gland, developed in Italy, originally followed in the footsteps of Kocher's technique [53, 54].

Some salient features of this technique include:

- Small incisions: Small incisions are made instead of a large incision, reducing discomfort and visible scarring.
- Improved visualization: The endoscopic camera provides an enlarged and detailed view of the surgical area, allowing for greater precision during surgery.
- Less damage to surrounding tissues: By minimizing the size of incisions, this technique aims to reduce damage to surrounding tissues and speed recovery time.
- This allows for improved cosmetic results, reduced postoperative pain, and no increase in complications, resulting in a shorter hospital stay [55].

These techniques can be performed through regular neck access or by moving the incisions away from the neck, known as remote access techniques.

Remote Access Techniques

These techniques require a longer duration of surgery and a higher cost, and the incidence of complications has not yet been evaluated in large series [56], but they allow scars to disappear from the neck, resulting in a better quality of life and a reduction in the psychological effects of scarring [57, 58].

These techniques are:

Transaxillary The transaxillary approach described in 2000 by Ikeda [59] was the first to be described and is currently less widely used. In the largest series (5000 cases) [60], the complication rate was:

- *Transient hypoparathyroidism in patients with total thyroidectomy was 48.1%, and permanent 1.3%.*
- *Transient dysphonia in 2.5% of cases.*
- *These data are difficult to reproduce in less experienced centers.*

Retroauricular The retroauricular approach described in 2011 by Terris [61], very rarely used in thyroid surgery at present, is indicated for lobectomies. In the larger series collected from several centers [62], the complication rates were:

- *3.9% of temporal recurrent paralysis.*
- *No cases of hypocalcemia (it should be remembered that lobectomies are performed via this approach).*
- *Hematomas, 2.9%.*

Trans Oral (TOETVA, TransOral Endoscopic Thyroidectomy—Vestibular Approach) described in 2010 by Wilhelm [63] via the floor of the mouth, perfected in 2017 via the vestibular approach [64]. In larger multi-center compilation series [65, 66], complication rates were:

- *Transient hypoparathyroidism was 7.65 to 10.9%.*
- *Transient recurrent paralysis 5.9%.*
- *Temporary hypocalcemia 4% and permanent hypocalcemia 0.8%.*
- *Seromas in 5%.*
- *Transient mentonian nerve injury in 1%.*

Bilateral Axillo-Breast Approach (BABA) Originally described by Youn in 2007 in its current form [67], it is mostly indicated for differentiated thyroid cancers with similar results to open surgery [68]. In the larger multi-center compilation series [69, 70] the complication rates were:

- *Transient hypoparathyroidism occurred in 14.2% and permanent hypoparathyroidism in 1.5%.*
- *Temporary recurrent paralysis occurred in 0.6% to 14.2% and permanent paralysis in 0% to 0.2%.*

Use of Robotics in Thyroid Surgery

Robotic surgery in thyroidectomy involves the use of robotic systems to perform the procedure, they are most commonly used in association with remote access [71].

Some associated features and benefits are:

- Improved accuracy: Robotic surgery allows for improved three-dimensional accuracy and greater surgeon dexterity, especially in difficult-to-access areas such as trans-axillary surgery.
- Small incisions: Although a robotic system is used, incisions remain small, contributing to an improved aesthetic appearance.
- Less bleeding: The precision of the robotic system can help reduce the risk of bleeding during surgery.
- Accelerated recovery: Although the evidence is evolving, some studies suggest that robotic surgery may be associated with faster recovery compared to conventional techniques.

Regarding the outcomes of robotic techniques with open or endoscopic techniques, in a meta-analysis [72] there was no difference in the incidence of complications, except for less hypoparathyroidism in BABA and in a comparative study of outcomes of open, endoscopic, and robotic thyroidectomy on 256 patients no difference in voice and swallowing, but more pain and discomfort in those operated by remote access [73].

Other Ablation Techniques

They are mainly used in benign nodular pathology and can be classified as follows:

- Ablation with energy: laser, radiofrequency, or microwave, the latter being the most commonly used at present.
- Chemical ablation: ethanol injection, used almost exclusively for cystic lesions.

It has a complication rate of around 7%, the most common complication being injury to the recurrent laryngeal nerves [62].

8 Outcomes and Effectiveness in Thyroid Gland Surgery

Evaluating the outcomes and effectiveness of thyroid gland surgery involves considering several aspects, including success rates, recurrence, and the patient's postoperative quality of life.

1. Success and recurrence rates

Success Rates Surgery of the thyroid gland, when properly performed, often has high success rates in the treatment of various conditions, such as benign thyroid nodules, thyroid cancer, and functional thyroid disorders. Success is assessed by complete removal of the target condition and restoration of adequate thyroid function.

Recurrence In thyroid cancer, thyroid cancer is carefully monitored for any signs of recurrence. Recurrence may occur in a small proportion of cases and may be due to the presence of residual cancer cells or the formation of new tumors. Follow-up strategies include blood tests, imaging studies, and, in some cases, additional treatments such as radioactive iodine therapy.

2. Postoperative Quality of Life Assessment

Thyroid Function After surgery, especially in cases of total thyroidectomy, it is crucial to maintain an adequate hormonal balance by administering synthetic thyroid hormones. This ensures that the patient does not experience symptoms of hypothyroidism.

Short-Term and Long-Term Complications Postoperative quality of life is affected by the presence or absence of complications. Short-term complications may include healing problems, temporary voice changes, and alterations in calcium levels. In the long term, proper management of thyroid function and prevention of recurrences contribute to a better quality of life.

Aesthetics and Vocal Function In surgeries involving incisions in visible areas, such as the neck, the aesthetic appearance of the scar may influence perceived quality of life. In addition, in cases of vocal impairment, vocal physiotherapy may be beneficial in improving vocal function.

Psychological Follow-Up Thyroid surgery, especially when cancer is involved, can have a significant emotional impact on patients. Psychological follow-up and support are important to address any anxiety or concerns related to the disease and treatment.

Comprehensive postoperative quality of life assessment focuses not only on the resolution of the underlying condition, but also on the patient's overall physical, emotional, and social well-being. Multidisciplinary management, including collaboration between surgeons, endocrinologists, psychologists, and other healthcare professionals, contributes to.

9 Ongoing Research and Emerging Areas of Study

Ongoing research in the field of thyroid gland surgery focuses on improving existing techniques, exploring new technologies, and addressing specific clinical challenges. Some areas of research and possible future developments are described here:

Personalized Surgery Genomic research is advancing towards personalized surgery, tailoring surgical and therapeutic approaches according to the unique genetic characteristics of each patient. Retroauricular access lymphadenectomy [74].

Advanced Robotic Surgery The evolution of robotic surgery could allow for even more precise and less invasive procedures. More sophisticated robotic systems are being investigated to improve surgeon dexterity and further reduce surgical trauma [75]. Ethical regulation for the development of autonomous robotic surgery has also been proposed [76].

Molecular Tests for Cancer Diagnosis In cytology with Bethesda thyroid nodule paaf II and IV with positive predictive values of 97% and negative predictive values of 96% [77].

Use of One-Step Nucleic Acid Amplification (OSNA) To identify lymph node metastases during surgery for papillary thyroid cancer with sensitivities of 87–89% and specificities of 86–94% [78].

Advanced Imaging Technologies The development of advanced imaging technologies, such as molecular imaging, could improve visualization during surgery and aid in identifying malignant tissue more accurately. Use of contrast-enhanced ultrasound to assess adequately treated areas with ablation or ultrasound elastography for diagnosis of cancer in nodules [10, 79].

Personalized Adjuvant Therapies Advances in adjuvant therapies, such as radioactive iodine therapy, aim to personalize treatments to the specific characteristics of each patient, minimizing side effects and improving efficacy. In cases of inadequate treatment of thyroid cancer with conventional therapies of surgery and radioactive iodine, new drugs are being developed based on the inhibition of signal-transmitting proteins inside the cell, known generically as "kinase inhibitors" [80]. They are of three types:

For differentiated cancer:

- *Multikinase inhibitors: Lenvatenib and Sorafenib.*
- *RET inhibitors: Selpercatinib and Pralsetinib.*
- *NTRK inhibitors: Larotrectinib and Entrectinib.*

For spinal cancer:

- *Multikinase inhibitors: Vandetanib and Cabozantinib.*
- *RET Inhibitors: Selpercatinib.*

For anaplastic cancer:

- *BRAF and MEK inhibitors: Dabrafenib and Trametinib.*
- *RET inhibitors: Selpercatinib.*
- *NTRK inhibitors: Larotrectinib and Entrectinib.*

Artificial Intelligence in Diagnostics

The application of artificial intelligence in image interpretation and preoperative diagnosis could increase the accuracy and speed of detection of thyroid disease [81–83].

10 Conclusions

Thyroid gland surgery is a dynamic field that has experienced remarkable advances in diagnosis, surgical techniques, and postoperative management. Some of the main findings and trends include:

Advances in Surgical Techniques Surgical interventions on the thyroid gland have evolved significantly, from conventional procedures to minimally invasive approaches and robotic-assisted techniques. These advances aim to improve precision, reduce surgical trauma, and speed patient recovery.

Personalized Approach Genomic research and targeted therapies are leading to a more personalized approach in the treatment of thyroid diseases, especially thyroid cancer. The identification of biomarkers and the understanding of the molecular characteristics of tumors allow for more targeted therapeutic strategies.

Impact on Quality of Life Postoperative quality of life is a crucial aspect that has gained importance in research. From managing thyroid function to addressing short- and long-term complications, the aim is to improve the overall patient experience.

Technological Developments The incorporation of advanced technologies, such as robotic surgery and artificial intelligence, promises to take thyroid surgery to new levels of diagnostic accuracy and efficacy.

Need for Continued Research Continued research in thyroid surgery is essential to further improve clinical outcomes and quality of life for patients. Current challenges, such as prevention of complications and optimization of thyroid function, require a multidisciplinary approach and collaboration between researchers, surgeons, and other healthcare professionals.

References

1. Filetti S, Durante C, Hartl D et al (2019) Thyroid cancer: ESMO clinical practice guidelines for diagnosis, treatment and follow-up. Ann Oncol 30:1856–1883. https://doi.org/10.1093/annonc/mdz400
2. Davies TF, Ando TA, Lin R-Y et al (2005) Thyrotropin-receptor-associated diseases: from adenomata to Graves' disease. J Clin Invest 115:1972–1983
3. Prabhakar BS, Bahn RS, Smith TJ (2003) Current perspective on the pathogenesis of Graves' disease and ophthalmopathy. Endocr Rev 24:802–835
4. Garmendia Madariaga A, Santos Palacios S, Guillén-Grima F et al (2014) The incidence and prevalence of thyroid dysfunction in Europe: a meta-analysis. J Clin Endocrinol Metab 99(3):923–931

5. Lucas A, Julian MT, Canton A et al (2010) Undiagnosed thyroid dysfunction, thyroid antibodies, and iodine excretion in a Mediterranean population. Endocrine 38:391–396
6. Vanderpump MP (2011) The epidemiology of thyroid disease. Br Med Bull 99 (1): 39–51. Disponible en: http://bmb.oxfordjournals.org/content/99/1/39.long
7. Surks MI, Ortiz E, Daniels GH, et al (2004) Subclinical thyroid disease: scientific review and guidelines for diagnosis and management. JAMA [Internet] 291(2):228–238. Disponible en: http://jama.jamanetwork.com/article.aspx?articleid=197994
8. Mazzaferri EL (1993) Management of a Solitary Thyroid Nodule. N Engl J Med [Internet] 328:553–559. Disponible en: http://www.nejm.org/doi/full/10.1056/NEJM1993022 53280807
9. Grussendorf M, Ruschenburg I, Brabant G (2022) Malignancy rates in thyroid nodules: a long-term cohort study of 17,592 patients. European Thyroid Journal 11:e220027. https://doi.org/10.1530/ETJ-22-0027
10. Durante C, Hegedüs L, Czarniecka A et al (2023) 2023 European thyroid association clinical practice guidelines for thyroid nodule management. European Thyroid Journal 12(5):e230067. https://doi.org/10.1530/ETJ-23-0067
11. European Network of Cancer Registries Factsheets (2017). https://www.encr.eu/sites/default/files/factsheets/ENCR_Factsheet_Thyroid_2017
12. Gallardo E, Medina J, Sánchez JC (2019) SEOM clinical guideline thyroid cancer. Clin Transl Oncol 2020(22):223–235. https://doi.org/10.1007/s12094-019-02284-8
13. Cameselle-Teijeiroa JM, Sobrinho-Simõesc M (2018) New WHO classification of thyroid tumors: a pragmatic categorization of thyroid gland neoplasms. Endocrinol Diabetes Nutr 65(3):133–135. https://doi.org/10.1016/j.endinu.2017.11.012
14. Cancer Genome Atlas Research Network (2014) Integrated genomic characterization of papillary thyroid carcinoma. Cell 159:676–690
15. Ibrahimpasic T, Xu B, Landa I et al (2017) Genomic alterations in fatal forms of non-anaplastic thyroid cancer: identification of MED12 and RBM10 as novel thyroid cancer genes associated with tumor virulence. Clin Cancer Res 23(19):5970–5980
16. Menéndez E, Santamaría J (2008) Grupo de trabajo de cáncer de tiroides de la SEEN. Guía para pacientes con cáncer diferenciado de tiroides. SEEN. Ediciones Mayo S.A, Madrid [actualizado 4 mar 2013; citado 19 may 2014]. Disponible en: http://www.seen.es/publico/enfermedades/tiroides.aspx
17. Agrawal N, Jiao Y, SausenM et al. (2013) Exomic sequencing of medullary thyroid cancer reveals dominant and mutually exclusive oncogenic mutations in RET and RAS. J Clin Endocrinol Metab; 98(2):E364–E369
18. Elisei R, Alevizaki M, Conte-Devolx B et al (2013) 2012 European thyroid association guidelines for genetic testing and its clinical consequences in medullary thyroid cancer. Eur Thyroid J 1:216–231
19. de Angulo Martín DR, Ruiz VM, Pérez PP (2010) Massive hemoperitoneum due to rupture of an spontaneous intramural oesophageal haematoma. Cir ESP 87(5):319–333. https://doi.org/10.1016/j.ciresp.2009.03.025
20. Papi G, Fadda G, Corsello SM et al (2007) Metastases to the thyroid gland: prevalence, clinicopathological aspects and prognosis: a 10 years experience. Clin Endocrinol 66:565–571
21. Moreira POO, Reyes MDO, De Jesús S, Recalde F et al (2022) Técnicas quirúrgicas en lobectomía parcial tiroidea en pacientes con hipertiroidismo primario. Riesgos y beneficios RECIMUNDO 6(2):291–299. https://recimundo.com/index.php/es/article/view/1573
22. Carcedo-Sañudo EG, de las Heras-Flórez P, Morales-Medina G et al (2021) Puntos clave en la cirugía de la glándula tiroides. Rev. ORL 12(4):359–370
23. Lorente Poch L (2020) Técnica quirúrgica de la tiroidectomía y vaciamientos ganglionares. En Joaquín Gómez Ramirez. 2020. Cirugía Endocrina. Guías clínicas de la Asociación española de cirujano 3ª ed. Aran ediciones. ISBN 978-84-18116-81-0
24. Pablo Moreno Llorente, Erick A. Gonzales Laguado et al (2021) Surgical approaches to thyroid. Cir Esp; 99 (4): 267–275
25. Durán Poveda, M (2020) Complicaciones posquirúrgicas en cirugía tiroidea. En Joaquín Gómez Ramirez. 2020. Cirugía Endocrina. Guías clínicas de la Asociación española de cirujano 3ª ed. Aran ediciones. ISBN 978-84-18116-81-0
26. Refoyo JLP (2010) Complicaciones de la cirugía tiroidea. Revista de la Sociedad Otorrinolaringológica de Castilla y León, Cantabria y La Rioja 1(4):1–152
27. Rosato L, Avenia N, Bernante P et al (2004) Complications of thyroid surgery: analysis of a multicentric study on 14, 934 patients

operated on in Italy over 5 years. World J Surg 28(3):271–276. https://doi.org/10.1007/s00268-003-6903-1

28. Cappellacci F, Canu GL, Piras S et al (2022) Technological innovations in thyroid cancer surgery. Oncologie 24(1):35–50. https://doi.org/10.32604/oncologie.2022.020864
29. Calò PG, Erdas E, Medas F et al (2013) Late bleeding after total thyroidectomy: report of two cases occurring 13 days after operation. Clin Med Insights Case Rep 6:165–170. https://doi.org/10.4137/CCRep.S13024
30. Calò PG, Pisano G, Piga G et al (2010) Postoperative hematomas after thyroid surgery. Incidence and risk factors in our experience. Ann Ital Chir 81(5):343–347
31. Materazzi G, Ambrosini CE, Fregoli L et al (2017) Prevention and management of bleeding in thyroid surgery. Gland Surg 6(5):510–515. https://doi.org/10.21037/gs.2017.06.14
32. Canu GL, Cappellacci F, Abdallah A et al (2023) Surgical Management of Indeterminate Thyroid Nodules across different world regions: results from a retrospective multicentric (the MAIN-NODE) study. Cancers 15:3996. https://doi.org/10.3390/cancers15153996
33. Angell TE, Maurer R, Wang Z et al (2019) A cohort análisis of clinical and ultrasound variables predicting cancer risk in 20,001 consecutive thyroid nodules. J Clin Endocrinol Metab 104:5665–5672. https://doi.org/10.1210/jc.2019-00664
34. Cipriani NA, White MG, Angelos P, Grogan RH (2018) Large cytologically benign thyroid nodules do not have high rates of malignancy or false-negative rates and clinical observation should be considered: a meta-analysis. Thyroid 28:1595–1608. https://doi.org/10.1089/thy.2018.0221
35. Cabanillas ME, McFadden DG, Durante C (2016) Thyroid cancer. Lancet 388(10061):2783–2795. https://doi.org/10.1016/S0140-6736(16)30172-6
36. Morales García D (2020) Bocio multinodular. . En Joaquín Gómez Ramirez. 2020. Cirugía Endocrina. Guías clínicas de la Asociación española de cirujano 3ª ed. Aran ediciones. ISBN 978-84-18116-81-0
37. Ocaña Wilhelmi LT, Glückmann Maldonado E, Villuendas Morales F et al (2020) Indicaciones quirúrgicas en patología tiroidea benigna. Cir Andal 31(3):217–223. https://doi.org/10.37351/2020313.2
38. Sundaresh V, Brito JP, Wang Z, Prokop LJ et al (2013) Comparative effectiveness of therapies for Graves' hyperthyroidism: a systematic review and network meta-analysis. J Clin Endocrinol Metab 98:3671–3677
39. Scerrino G, Richiusa P, Graceffa G et al (2022) N.C. Editorial: recent advances in thyroid surgery. J Clin Med 11:7233. https://doi.org/10.3390/jcm11237233
40. Amit M, Binenbaum Y, Cohen JT et al (2013) Effectiveness of an oxidized cellulose patch hemostatic agent in thyroid surgery: a prospective, randomized, controlled study. J Am Coll Surg 217:221–225
41. Testini M, Marzaioli R, Lissidini G et al (2009) The effectiveness of FloSeal matrix hemostatic agent in thyroid surgery: a prospective, randomized, control study. Langenbeck's Arch Surg 394:837–842
42. Scerrino G, Paladino NC, Di Paola V et al (2013) The use of haemostatic agents inthyroid surgery: efficacy and further advantages. Collagen-fibrinogen-thrombin patch (CFTP) versus cellulose gauze. Ann Ital Chir 84:545–550
43. Khadra H, Bakeer M, Hauch A et al (2018) Hemostatic agent use in thyroid surgery: a metaanalysis. Gland Surg 7(Suppl 1):S34–S41. https://doi.org/10.21037/gs.2018.03.02
44. Papavramidis TS, Pliakos I, Chorti A et al (2020) Comparing LigasureTM exact dissector with other energy devices in total thyroidectomy: a pilot study. Gland Surg 9(2):271–277. https://doi.org/10.21037/gs.2020.02.05
45. Paladino NC, Guérin C, Graziani J et al (2021) Predicting risk factors of postoperative hypocalcemia after total thyroidectomy: is safe discharge without supplementation possible? A large cohort study. Langenbeck's Arch Surg 406:2425–2431
46. Lu W, Chen Q, Zhang P et al (2022) Near-infrared autofluorescence imaging in thyroid surgery: a systematic review and meta-analysis. J Investig Surg 35:1723–1732
47. Choi JY, Yu HW, Bae IE, Kim et al (2018) Novel method to save the parathyroid gland during thyroidectomy: subcapsular saline injection. Head Neck 40(4):801–807. https://doi.org/10.1002/hed.25068
48. Lahey FH, Hoover WB (1938) Injuries to the recurrent laryngeal nerve in thyroid operations their management and avoidance. Ann Surg 108(4):545–562. https://doi.org/10.1097/00000658-193810000-00006
49. Wu CW, Dionigi G, Barczynski M et al (2018) International neuromonitoring study group guidelines 2018: part II: optimal recurrent

laryngeal nerve management for invasive thyroid cancerincorporation of surgical, laryngeal, and neural electrophysiologic data. Laryngoscope 128(Suppl 3):S18–S27. https://doi.org/10.1002/lary.27360

50. Schneider R, Randolph GW, Barczynski M et al (2016) Continuous intraoperative neural monitoring of the recurrent nerves in thyroid surgery: a quantum leap in technology. Gland Surg 5:607–616
51. Calò PG, Medas F, Conzo G et al (2017) Intraoperative neuromonitoring in thyroid surgery: is the two-staged thyroidectomy justified? Int J Surg 41(Suppl 1):S13–S20. https://doi.org/10.1016/j.ijsu.2017.02.001
52. Sitges-Serra A, Fontané J, Dueñas JP et al (2013) Prospective study on loss of signal on the first side during neuromonitoring of the recurrent laryngeal nerve in total thyroidectomy. Br J Surg 100(5):662–666. https://doi.org/10.1002/bjs.9044
53. Lage Fernández FJ, Paulos Novoa M, Parente AP (2021) Minimally invasive video-assisted thyroidectomy (MIVAT): how to do it (with video). Eur Ann Otorhinolaryngol Head Neck Dis S1879–7296(21):00155–00151. https://doi.org/10.1016/j.anorl.2021.07.004
54. Dionigi G, Dralle H, Materazzi G et al (2020) 20th birthday to minimally invasive video-assisted thyroidectomy. J Endocrinol Investig 43:385–388
55. Scerrino G, Melfa G, Raspanti C et al (2019) Minimally invasive video-assisted thyroidectomy: analysis of complications from a systematic review. Surg Innov 26:381–387
56. Graves CE, Suh I (2020) The current status of remote access thyroidectomy in the United States. Surgery 168:845–850
57. Goswami S, Peipert BJ, Mongelli MN et al (2019) Clinical factors associated with worse quality-of-life scores in United States thyroid cancersurvivors. Surgery 166:69–74
58. Best AR, Shipchandler TZ, Cordes SR (2017) Midcervical scar satisfaction in thyroidectomy patients. Laryngoscope 127:1247–1125
59. Ikeda Y, Takami H, Sasaki Y et al (2000) Endoscopic neck surgery by the axillary approach. J Am Coll Surg 191:336–340
60. Kim MJ, Nam KH, Lee SG et al (2018) Experience of 5000 gasless Transaxillary robotic thyroidectomies. World J Surg 42(2):393–401. https://doi.org/10.1007/s00268-017-4209-y
61. Terris DJ, Singer MC, Seybt MW (2011) Robotic facelift thyroidectomy: II. Clinical feasibility and safety. Laryngoscope 121:1636–1641
62. Duke WS, Holsinger FC, Kandil E et al (2017) Remote access robotic facelift thyroidectomy: a multi-institutional experience. World J Surg 41(1):116–121. https://doi.org/10.1007/s00268-016-3738-0
63. Wilhelm T, Metzig A (2010) Video. Endoscopic minimally invasive thyroidectomy: first clinical experience. Surg Endosc 24:1757–1758
64. Anuwong A (2016) Transoral endoscopic thyroidectomy vestibular approach: a series of the first 60 human cases. World J Surg 40(3):491–497
65. Anuwong A, Ketwong K, Jitpratoom P et al (2018) Safety and outcomes of the Transoral endoscopic thyroidectomy vestibular approach. JAMA Surg 153:21–27
66. Lira RB, De Cicco R, Rangel LG et al (2021) Transoral endoscopic thyroidectomy vestibular approach: experience from a multicenter national group with 412 patients. Head Neck. https://doi.org/10.1002/hed.26846
67. Choe JH, Kim SW, Chung KW et al (2007) Endoscopic thyroidectomy using a new bilateral axillo-breast approach. World J Surg 31:601–606
68. Lang BH, Wong CK, Tsang JS et al (2015) A systematic review and meta-analysis evaluating completeness and outcomes of robotic thyroidectomy. Laryngoscope 125(2):509–518. https://doi.org/10.1002/lary.24946
69. Lee KE, Kim E, do Koo H, Choi JY et al (2013) Robotic thyroidectomy by bilateral axillo-breast approach: review of 1,026 cases and surgical completeness. Surg Endosc 27:2955–2962
70. You JY, Kim HK, Kim HY et al (2021) Bilateral axillo-breast approach robotic thyroidectomy: review of a single surgeon's consecutive 317 cases. Gland Surg 10(6):1962–1970. https://doi.org/10.21037/gs-21-50
71. Granell J (2022) Actualización en cirugía Tiroidea y paratiroidea de acceso remoto. Rev ORL 13(2):159–169. https://doi.org/10.14201/orl.27167
72. Xing Z, Qiu Y, Abuduwaili M et al (2021) Surgical outcomes of different approaches in robotic assisted thyroidectomy for thyroid cancer: a systematic review and Bayesian network meta-analysis. Int J Surg 89:105941. https://doi.org/10.1016/j.ijsu.2021.105941
73. Ha TK, Kim DW, Park HK et al (2018) Comparison of postoperative neck pain and discomfort, swallowing difficulty, and voice change after conventional open, endoscopic, and robotic thyroidectomy: a single-center cohort

study. Front Endocrinol (Lausanne) 9:416. https://doi.org/10.3389/fendo.2018.00416

74. Greer Albergotti W, Kenneth Byrd J, De Almeida JR et al (2014) Robot-assisted level II-IV neck dissection through a modified face-lift incision: initial north American experience. Int J Med Robot 10(4):391–396. https://doi.org/10.1002/rcs.1585
75. Park YM, Choi EC, Kim SH et al (2021) Recent progress of robotic head and neck surgery using a flexible single port robotic system. J Robot Surg. https://doi.org/10.1007/s11701-021-01221-8
76. O'Sullivan S, Nevejans N, Allen C et al (2019) Legal, regulatory, and ethical frameworks for development of standards in artificial intelligence (AI) and autonomous robotic surgery. Int J Med Robot 15:e1968
77. Roth MY, Witt RL, Steward DL (2018) Molecular testing for thyroid nodules: review and current state. Cancer 124:888–898
78. Tranoulis A, Georgiou D, Yap J et al (2021) The evolving role of one-step nucleic acid amplification (OSNA) for the intraoperative detection of lymph node metastases: a diagnostic accuracy meta-analysis. Eur J Surg Oncol 47(6):1233–1243. https://doi.org/10.1016/j.ejso.2020.12.001
79. Cantisani V, De Silvestri A, Scotti V et al (2022) US-Elastography with different techniques for thyroid nodule characterization: systematic review and meta-analysis. Front Oncol 12: 845549
80. National Comprehensive Cancer Network. NCCN Clinical Practice Guidelines in Oncology: Thyroid Carcinoma. V.2.2023 y American Cancer Society. Cancer Facts & Figures 2023. American Cancer Society, Atlanta
81. Zhao C-K, Ren T-T, Yin Y-F et al (2021) A comparative analysis of two machine learning-based diagnostic patterns with thyroid imaging reporting and data system for thyroid nodules: diagnostic performance and unnecessary biopsy rate. Thyroid 31:470–481
82. Park V, Han K, Seong YK et al (2019) Diagnosis of thyroid nodules: performance of a deep learning convolutional neural network model vs. Radiologists Sci Rep 9:17843
83. Kim W, Lee JK, Yu HW et al (2022) Advancements in thyroidectomy: a mini review. Endocrine 3:789–800. https://doi.org/10.3390/endocrines3040065

Chapter 15

Immunohistochemical Study Protocol in Thyroid Pathology

Nicole Dimas Vasquez, Tania Sierra Hompanera, and Montserrat Chao-Crecente

Abstract

Immunohistochemical study proves valuable in the diagnosis of neoplastic pathology. Its application allows us to differentiate the origin of tumors, as seen with thyroglobulin and TTF1 in well-differentiated tumors of follicular origin and calcitonin or other neuroendocrine markers for tumors arising from C cells. At times, immunohistochemical study becomes necessary for the classification of thyroid carcinomas arising from follicular cells, although specific antibodies that would enable the exclusion of malignancy or differentiation between follicular carcinomas and papillary carcinomas have generally not been found; hence, various antibody panels are utilized. HMBE1 and galectin 3 can be beneficial in distinguishing malignancy from benignity, while CK19, galectin 3, and HMBE1 aid in identifying papillary carcinomas of follicular variant. In the differential diagnosis depending on histological morphology, various panels with automated protocols prove greatly useful.

Key words Immunohistochemistry, Ommunocytochemistry, Antibodies, Staining, Thyroid neoplastic pathology

1 Introduction

Immunohistochemical study in thyroid pathology can prove to be a highly useful tool for the differential diagnosis of neoplastic pathology. At times, the performance of antibodies on paraffin samples and cytological samples from fine-needle aspiration proves to be greatly useful.

In the latest WHO classification, new considerations regarding low-risk neoplasms have been taken into account [1] (Table 1).

The importance of studying molecular alterations in different histological groups leads to the creation of new groups, such as the encapsulated follicular variant of papillary thyroid carcinoma (EFVPTC), which demonstrates that its biological behavior is more similar to the group of follicular carcinomas due to its ras mutations.

Constanza Contreras-Jurado (ed.), *Thyroid Hormones: Methods and Protocols*, Methods in Molecular Biology, vol. 2876, https://doi.org/10.1007/978-1-0716-4252-8_15,

Table 1
WHO classification of thyroid tumours

Follicular cell-derived neoplasms	
	(a) Thyroid follicular nodular disease
1. Bening tumors	(b) Follicular adenoma
	(c) Follicular adenoma with papillaryy architecture
	(d) Oncocytic adenoma of the tyroid
	(a) Non-invasive follicular thyroid neoplam with papillary-like nuclear features
2. Low-risk neoplasms	(b) Thyroid tumors of uncertain malignant potential
	(c). Hyalinizing trabecular tumor
3. Malignant neoplams	(a). Follicular thyroid carcinoma
	(b) Invasive encapsulated follicular variant papillary carcinoma
	(c) Papillary thyroid carcinoma
	(d) Oncocytic carcinoma of the thyroid
	(e) Follicular-derived carcinomas, high-grade
	Differentiated high-grade thyroid carcinoma
	Poorly differentiated thyroid carcinoma
	(f) Anaplastic follicular cell-derived thyroid carcinoma
Thyroid C-cell-derived carcinoma	
1. Medullary thyroid carcinoma	
Mixed medullary and follicular cell-derived carcinomas	
Silvary gland-type carcinomas of the thyroid	
1. Mucoepidermoid carcinoma of the thyroid	
2. Secretory carcinoma of salivary gland type	
Thyroid tumors of uncertain histogenesis	
1. Sclerosing mucoepidermoid carcinoma with eosinophilia	
2. Cribriform morular thyroid carcinoma	
Thymic tumors within the thyroid	
1. Thymoma family	
2. Spindle epithelial tumor with thymus-like elements	
3. Thymic carcinoma family	
Embryonal thyroid neoplams	
1. Thyroblastoma	

On the other hand, the term "microcarcinoma papillare" traditionally used for papillary carcinomas smaller than 1 cm is eliminated. However, it has been observed that there are "microcarcinomas" with aggressive characteristics. Within well-differentiated carcinomas, follicular and papillary carcinomas, a high-grade group is created based on the number of mitoses and necrosis.

Frequently, the use of immunohistochemical techniques for differential diagnosis is necessary to determine:

- Cellular origin of primary tumors: neuroendocrine tumors and follicular cell tumors.
- Classification of neoplasms of follicular origin.
- Differentiation between primary and metastatic tumors.
- Thyroid tumors of uncertain malignant potential.

1.1 Antibodies in Normal and Neoplastic Thyroid Parenchyma

The thyroid gland consists of an epithelial component of follicular cells and C cells, which are more difficult to visualize in routine hematoxylin-eosin sections.

Immunoreactive antibodies in normal thyroid:

Thyroglobulin It is a glycoprotein of 660 kD expressed in follicular cells. The staining intensity varies depending on whether it is atrophic cells distended by colloid or hyperplastic cells. The staining intensity is higher in hyperplastic pathologies such as Graves' disease or the hyperplastic phase of Hashimoto's thyroiditis.

Follicular adenomas are positive for thyroglobulin depending on their activity. They will present a higher intensity if they are hyperfunctioning and lower intensity in solid or oncocytic adenomas [2].

TTF1 It is a protein containing a homeodomain of the thyroid transcription factor. Its function is to regulate the expression of thyroperoxidase and the TGB gene in the thyroid [3].

PAX8 It is an essential transcription factor in thyroid organogenesis. Its expression has been primarily reported in the thyroid but also in renal neoplasms, bladder, and malignant tumors of Müllerian origin including the ovary [4, 5].

Strong nuclear staining has been reported in thyroid follicular cells, Müllerian epithelium, renal tubular epithelium, vas deferens epithelium, and pancreatic islet cells [6].

Thyroperoxidase is a thyroid-specific enzyme expressed diffusely in follicular cells [7]. The expression of thyroperoxidase is lost as dedifferentiation occurs in follicular cells, thus its loss can be used as a marker of malignancy.

Calcitonin It is the primary hormone secreted by C cells, which is involved in calcium metabolism along with parathyroid hormone secreted by the parathyroid glands [8].

Antibodies in neoplastic thyroid pathology

Thyroid carcinomas, arising from follicular cells, will exhibit positivity depending on their degree of differentiation.

Thyroglobulin Is an immunohistochemical marker found in 95% of papillary and follicular carcinomas, making it useful in the differential diagnosis with metastatic carcinomas [9].

The staining pattern may be patchy; therefore this should be considered in small biopsies, and its negativity does not exclude the diagnosis.

Anaplastic thyroid carcinomas typically test negative for thyroglobulin. It was observed that they could be focal positives in 15.6% of the cases, shown by a study which was reported by Ordoñez and coworkers [10]. Nonetheless, it was interpreted that the positive cells may represent trapped cells.

TTF1 has been observed to be expressed in 96% of papillary carcinomas, 100% of follicular carcinomas, 20% of oncocytic carcinomas, 100% of insular carcinomas, and 90% of medullary carcinomas [11]. It is a useful marker in differentiating primary thyroid tumors arising from follicular and parafollicular cells, but it is not positive in anaplastic carcinomas.

In the lung, it plays an important role in the expression of surfactant proteins and Clara cells. For this reason, it is a marker frequently used in lung carcinomas. It is expressed in 72% of lung adenocarcinomas and 10% of squamous cell carcinomas, and it is also expressed in small cell lung carcinomas. However, TTF1 expression has also been reported in extrapulmonary carcinomas such as prostate, bladder, and uterine cervix.

TTF1, TTF2, and PAX 8 are expressed in differentiated thyroid carcinomas. PAX 8 is expressed in 80% of anaplastic carcinomas [1]. TTF2 was negative in other neoplasms and non-thyroidal tissues.

CK19 is a low molecular weight cytokeratin found in various glandular epithelia and simple epithelia. It is normally not expressed in normal follicular epithelium, but focal expression has been described, especially in the presence of thyroiditis. Many studies have demonstrated positivity in 100% of papillary carcinomas [12]. Focal positivity is observed in adenomas and follicular carcinomas.

HMBE-1 is a monoclonal antibody that recognizes an unspecified antigen of microvilli in mesothelial cells, tracheal epithelium, and adenocarcinomas of the pancreas, lung, and breast. It is used for the differential diagnosis of benign and malignant thyroid lesions. Positivity was observed in 100% of papillary and follicular carcinomas, and focal positivity was observed in only 30% of benign lesions [13].

Galectin-3 is a beta-galactoside-binding lectin expressed in a wide range of normal and neoplastic tissues. Cellular proliferations are associated with an increase in the phosphorylated and non-phosphorylated forms of Galectin-3, expressed in the nucleus and cytoplasm respectively. Nuclear expression versus cytoplasmic expression are associated with neoplastic progression [14]. Galectin-3 has been shown to be a useful marker in the differential diagnosis of well-differentiated malignant thyroid lesions such as papillary carcinoma and the follicular variant of papillary carcinoma from benign lesions.

In various studies, it has been concluded that Galectin-3 could be useful in distinguishing benign and malignant thyroid neoplasms with a sensitivity of 99%, specificity of 98%, and positive predictive value of 92% [15].

TROP2 (Trophoblastic Cell Surface Antigen 2) is a transmembrane glycoprotein associated with neoplastic progression in a wide variety of epithelial carcinomas such as ovarian, colorectal, pancreatic, gastric, pulmonary, endometrial, and squamous cell carcinoma of the oral cavity [16]. Many studies have shown that TROP2 expression is associated with aggressiveness and worse prognosis. In a study by Haiyan Liu and colleagues, TROP2 was demonstrated as a potential marker that allows differentiation between follicular and classic variants of papillary thyroid carcinoma [3].

PAX8 (Paired Box Gene 8) is a marker that shows positivity in thyroid neoplasms, but it is important to note that it can also be positive in the thymus and parathyroid glands. Among thyroid tumors, a majority of non-follicular tumors are negative, and positivity in medullary carcinomas is uncommon. The monoclonal antibody is more specific for thyroid tissue [17]. It is one of the main markers for determining the thyroid origin of follicular thyroid carcinoma alongside TTF1 and thyroglobulin [18].

CITED1 is a nuclear protein involved in the co-regulation of transcription factors. This protein is present in different types of papillary carcinomas (87% to 93%). However, it is also present in 50% of follicular carcinomas, 16% of follicular adenomas, and 24%

of goiters, making it a marker with low sensitivity and specificity [19].

MIB1 is a marker of cell proliferation that proves useful when it exhibits a membranous pattern for diagnosing hyalinizing trabecular tumor. This pattern can be observed at room temperature due to cross-reactivity with an epitope present on the plasma membrane [20].

1.2 Molecular Immunohistochemical Markers

The molecular understanding of thyroid neoplasms has allowed for their reclassification based on their aggressiveness and biological behavior, which also determines therapeutic decisions.

Current immunohistochemical studies enable the detection of mutations in BRAF V600E (VE1 clone), RAS Q61R mutations (SP147 clone), and rearrangements in ALK and NTRK, as seen in other pathologies. Although the use of immunohistochemistry is only approved for BRAF for clinical management. The reported sensitivity for BRAF pV600E in cellular block is 94 to 100%. This method is utilized in some services after obtaining a fine-needle aspiration biopsy to rule out papillary carcinoma-associated malignancy, although it is not recommended for monolayer samples [21].

The primary antibodies commonly used to determine the primary origin are usually thyroglobulin for neoplasms derived from follicular cells and calcitonin for those derived from parafollicular cells. Although the classification of primary thyroid neoplasms is based on morphological characteristics, sometimes other antibodies such as TTF1 and PAX8 need to be added. In the differential diagnosis of follicular variant papillary carcinomas, follicular adenomas, and follicular carcinomas, CK19, Galectin-3, and HMBE1 can also be included among others. The combination of these markers can aid in diagnosis, considering that individually none of them has high specificity.

Other immunohistochemical techniques are also useful in determining the presence of vascular invasion, especially for the diagnosis of angioinvasive follicular carcinomas. The most commonly used antibodies are CD31, Factor VIII, and CD34 for endothelial cells and D2–40 (podoplanin) for lymphatic vessels. Histochemical techniques can also be employed to visualize the elastic fibers of vessel walls [21, 22]. (Figs.1, 2, 3, 4, and 5).

2 Materials

2.1 Management of Surgical Specimens

- Fixative agent 10% formaldehyde.

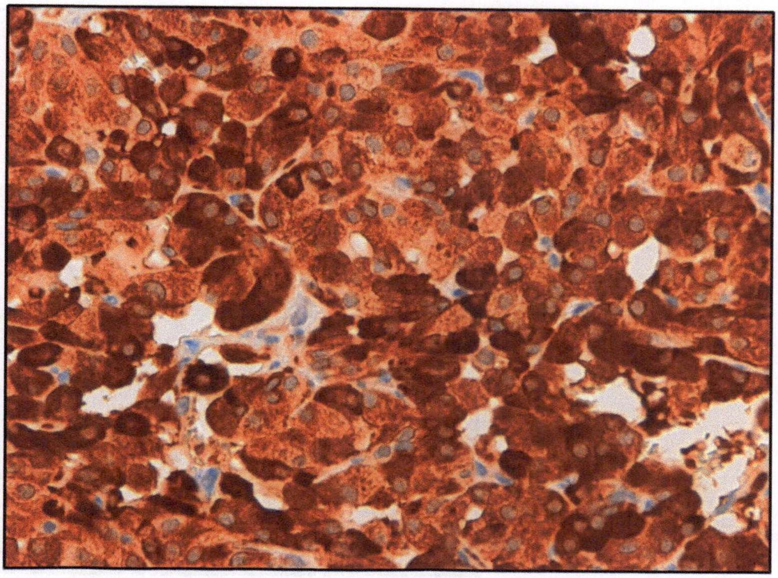

Fig. 1 Cromogranin. Granular cytoplasmic staining in neoplastic cells of a medullary carcinoma

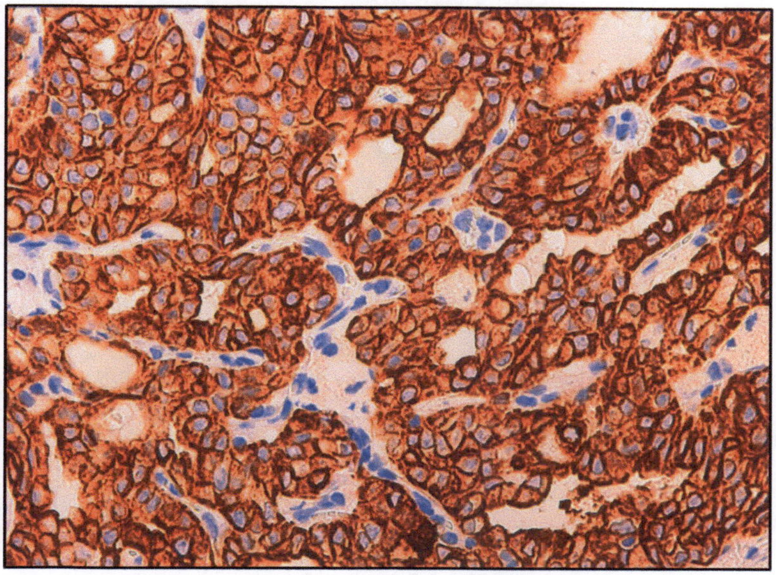

Fig. 2 Cytokeratin 19. Cytoplasmic staining

2.2 Inmunohistochemical Solutions

- Wash Buffer: a concentrated buffer solution that requires initial dilution (1:10). The diluted solution is used for washing between steps.

  Composition: Buffered saline solution with TRIS, surfactant, and 3.5% ProClin 950.

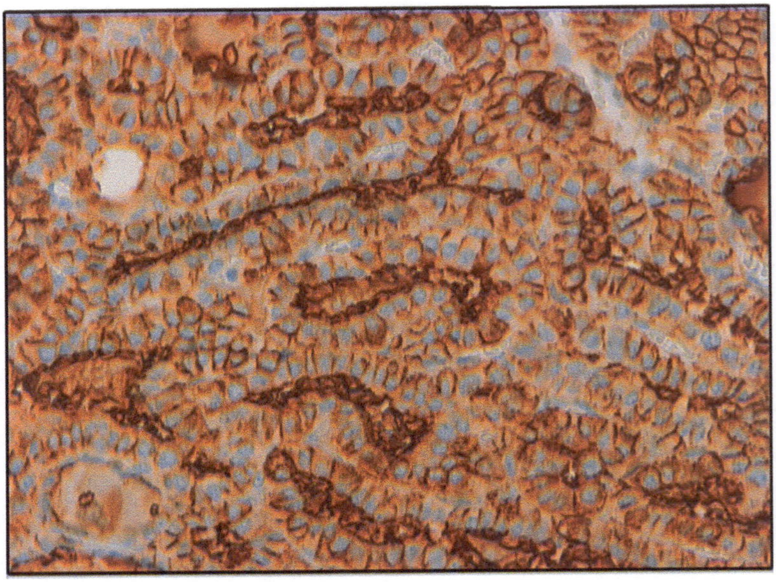

Fig. 3 HMBE1. Cytoplasmic staining

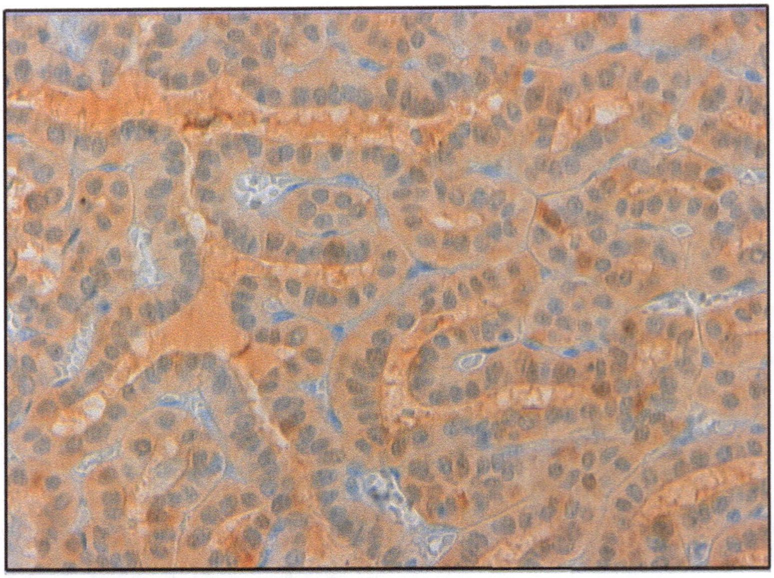

Fig. 4 Galectina. Cytoplasmic staining

- Dewax Solution: to remove excess paraffin before staining. Composition: solvent-based solution.
- Epitope retrieval 1 (pH 6) or 2 (pH 9): depending on the antibody to be detected.

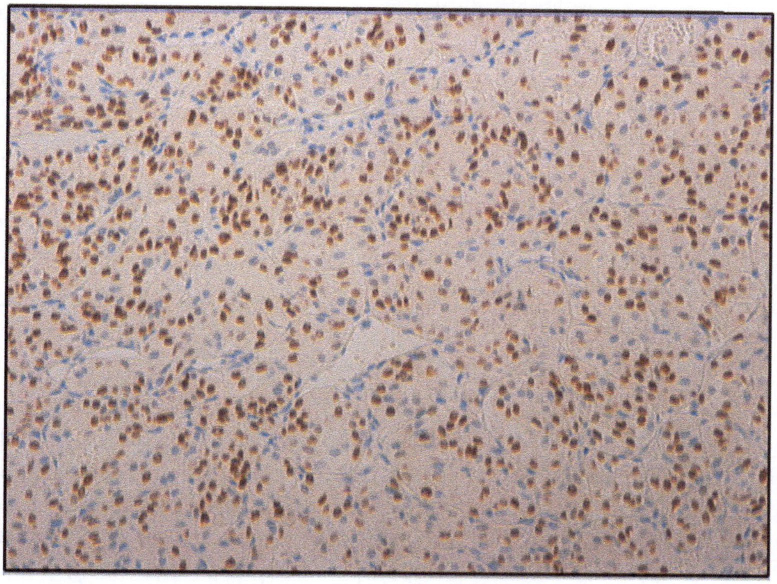

Fig. 5 TTF1. Nuclear staining

Composition: EDTA-based buffer with surfactant.

2.3 Immunocytochemical Solutions

- Diff-Quik.
  - (a) Fixative solution: triarylmethane dye and methanol.
  - (b) Eosinophilic staining solution: eosin Y.
  - (c) Basophilic staining solution: atrazine dye, methylene chloride, and azure A.
- Cytolyt.
  - (a) Composition: methanol and water.
- PreservCyt.
  - (b) Composition: methanol, distilled water, sodium EDTA, cholic acid, and glacial acetic acid.

3 Methods

Thyroid nodules suspected of malignancy are studied in detail histologically from a hemithyroidectomy and/or by fine needle aspiration-puncture (FNA) to obtain cytological material.

The processing of surgical specimens in pathology services is a fundamental phase for histopathological diagnosis. The quality of surgical specimen processing is essential as it directly influences the viability and preservation of tissue structures, thereby ensuring an adequate sample for the application of immunohistochemical techniques aimed at identifying and characterizing specific markers with clinical and diagnostic relevance.

3.1 Immunohistochemical Study

3.1.1 Management of Surgical Specimens

1. Sample Reception: The surgical specimen is received, naming the surgical procedure performed, whether total or partial thyroidectomy, and specifying the laterality. Additionally, the sample reception, whether fresh or fixed, is described.
2. Registration and Labeling: Each specimen is registered and labeled with relevant patient information (medical record number, patient name, specimen number, and date, among others).
3. Fixation: The specimen must be placed in a container with 10% formaldehyde, completely covering the surgical piece. If the specimen is very large, cuts should be made every centimeter, and it should be left to fix for 24 to 48 h. The function of 10% formaldehyde is to preserve the cellular structure, preventing decomposition and facilitating subsequent handling of the specimen.
4. Dimension and Description: The specimen is measured in three dimensions and weighed. Subsequently, external description is provided, including whether it is fragmented or complete, shape, symmetry, color, consistency, and presence of extrathyroidal tissues (Fig. 6).

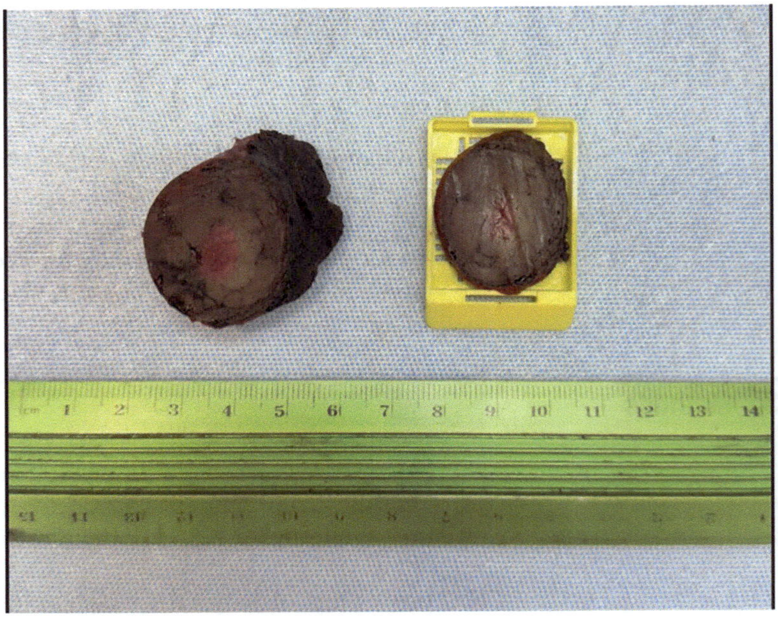

Fig. 6 Surgical specimen processing, specimen sectioning, and embedding in blocks

5. Specimen Marking facilitates histological localization of lesions and allows evaluation of margin involvement (*see* **Note 1**).
6. Specimen Sectioning: Serial and transverse cuts of approximately 2–3 mm thickness are made, evaluating thyroid tissue characteristics and describing them. If a single lesion or multiple lesions are evident, they should be measured and described. Sections of suspicious nodules should be perpendicular to the nodule capsule to reduce artifact.
7. Tissue Selection: The most representative areas for study and diagnosis are selected, such as areas of tumor capsule thickening, suspected invasion zones, and adjacent thyroid tissue for the likelihood of detecting other unseen carcinoma foci. These selected sections are placed in marked and numbered containers for each patient. The diameter and length of the selected tissue depend on the capsules used in each laboratory. The number of capsules is important and may vary depending on the suspected pathology, although it is estimated that approximately 97% of carcinomas can be detected with 10 blocks [23].

3.1.2 Sample Processing

1. Sample dehydration: It is a critical process for tissue preparation, in which, after fixation with formaldehyde (1 h), the sample is placed in a series of alcohol solutions with increasing concentrations (70%, 80%, 90%, 95%, and 100%) to remove the water (each step lasts from 30 min to 1 h). Subsequently, clearing is done with xylene to remove the alcohol, generally with 2 to 3 changes of xylene (each change lasts from 30 min to 1 h). Finally, the tissue sample is transferred to liquid (melted) paraffin.
2. Sectioning: The paraffin-embedded section is cut using a microtome, generating serial sections approximately 3–5 micrometers thick. These sections are mounted on glass slides for subsequent analysis.
3. Slides are placed in an oven at 56–58 °C for 15–20 min to remove paraffin.

3.1.3 Immunohistochemistry Protocol

Cut serial sections from the selected block at 3 microns. On the same slide, a control (usually positive) for each immunohistochemistry technique to be performed is placed (*see* **Note 2**).

Sections are stained with special dyes that highlight cellular characteristics and facilitate visualization under the microscope:

1. Hematoxylin-eosin staining: Before performing any immunohistochemistry technique, it is necessary an initial diagnosis by hematoxylin-eosin. This staining is used for selected subsequent immunohistochemistry techniques.

   - Steps:

     Oven for 30 min.

     2 jars × 10 min in xylene.

     2 jars × 10 min in 100% ethanol.

     10 min in 96% ethanol.

     10 min in 80% ethanol.

     5 min in H2O.

     5–10 min in Carazzi's hematoxylin.

     15 min in running water.

     2 jars × 1 min in distilled water.

     1–2 min in alcoholic eosin.

     30 s in 70% ethanol.

     30 s in 96% ethanol.

     2 jars × 3 min in 100% ethanol.

     2 jars × 10 min in xylene.

     Mount with mounting medium and cover with a coverslip.

2. Antibodies staining.

   Select the most appropriate panel to confirm the diagnosis, subclassify, or provide prognostic or predictive information for treatment (*see* **Note 3**).

   - The sample is incubated with the selected marker for 20 min (*see* **Note 4**).
   - Three washes are performed with Wash Buffer solution.
   - The sample is incubated with hydrogen peroxide to neutralize the activity of the tissue's endogenous peroxidase and eliminate background for 8 min.
   - Three washes are performed with Wash Buffer solution.
   - A specific primary antibody is applied, in this case, the selected specific antibodies.
   - The Post-primary IgG binding reagent is added to localize mouse antibodies for 8 min.
   - Three washes are performed with Wash Buffer solution for 2 min each to remove nonspecific background.
   - The Poly-HRP IgG reagent (polymer) is added to localize existing rabbit antibodies from the previous step for 8 min.
   - The slide is washed with Wash Buffer solution for 2 min.

- The slide is washed with Wash Buffer solution for 2 min.
- The slide is washed with Wash Buffer solution for 1 min.
- Two more washes are performed with Wash Buffer solution.
- One wash is performed with distilled water.
- The chromogenic substrate (diaminobenzidine or DAB) is added for 10 min to visualize the previously created complex by oxidizing the peroxidase molecule of the polymer, resulting in a brown precipitate visible under the microscope.
- Four washes are performed with distilled water.
- Hematoxylin is added for 10 min for counterstaining (cellular nuclei of the tissue can be visualized in blue).
- One wash is performed with distilled water.
- Two washes are performed with Wash Buffer solution.
- One wash is performed with distilled water.

3.1.4 Mounting for Visualization

1. Once the immunostaining process is complete, the slides are placed in a basket for subsequent dehydration, undergoing successive baths in alcohols of increasing concentration, 5 min each (80%, 96%, 100%, and xylene).
2. Mounting for visualization under the microscope, the mounting is done with DPX, which is composed of xylene and ethylbenzene.

3.2 Immunocytochemical-Study

The protocol for preparing samples for microscopy includes the following steps:

3.2.1 Sample Processing from Fine Needle Aspirations of the Thyroid

1. Spread and fixation of samples:

   This can be done in two ways:

   - Spreading and fixation of the material are performed directly on the slide (FNAB), immediately after obtaining the sample. Air-dried slides for subsequent staining with Diff-Quik, and crystals fixed in 96% alcohol (for Papanicolaou staining).
   - Needle wash vial that typically arrives at the laboratory in the form of liquid fluid which must be processed in ThinPrep, automatically concentrates the cell population on the slide for subsequent staining and assessment under the microscope. For spreading perform the following steps:
     1. Collect thyroid fluid using FNAB in 30 mL of Cyto-Lyt solution (alcohol-based solution).

2. Concentrate the cell population by centrifugation: 1800 rpm for 10 min (*see* **Note 5**) or 2800 rpm for 5 min.
3. Discard the supernatant and resuspend the cell pellet using a vortex or plastic pipette.
4. Assess the appearance of the cell pellet. If the cell pellet is not free of blood, add 30 mL of CytoLyt and repeat **step 2**.
5. Add the sample to a PreservCyt vial (that allows transportation and cell preservation) and let it stand for 5 min.
6. Transfers cells to a slide, providing a thin and uniform cell layer.

2. Once the processing is complete, the obtained slides are placed in racks (with 96% alcohol solution) ready for subsequent Papanicolaou staining.

3.2.2 Staining

For cytology staining, the most commonly used techniques are:

1. **Papanicolaou:** It utilizes hematoxylin as the nuclear stain, typically Harris hematoxylin, and various mixtures of cytoplasmic stains such as orange G, eosin, light green SF, and Bismarck brown (EA mixture). The solutions are stored in a cool place at room temperature. The staining time for each dye varies according to the commercial brand, concentration, and filtration of the reagents.

   Protocol:

   - Fix the sample to 96% alcohol.
   - Successively immerse in 80% alcohol, 70% alcohol, 50% alcohol, and water for 1 min each.
   - Stain with Harris Hematoxylin for approximately 5 min.
   - Immerse in distilled water 6 times for 1 second each.
   - Immerse in 0.5% Hydrochloric Acid, 8 times for 1 second each.
   - Wash with running water for 5 min, then pass the sample through increasing gradient alcohols, 50%, 70%, 80%, and 96%, for 30 s each.
   - Stain with Orange G solution for 1 to 1.5 min.
   - Remove excess dye with two baths of 96% alcohol, dipping the preparation 2 times in each bath for 3 to 4 s.
   - Stain with EA 50 solution for 1.5 to 2 min.
   - Wash in 3 different containers with 96% alcohol, dipping the preparation 2 times for 3 to 4 s each.

- Wash in absolute alcohol for 30 s.
  - Immerse the preparation for 3 min in 2 containers with Xylene baths.
  - Mount the preparation with a mounting medium.
  - Observe under the microscope.

2. **Diff-Quik:** It is a rapid staining method whose reagents are commercially available. It consists of 3 liquids:

   first Solution: fixative with methanol.

   second Solution: cytoplasmic stain.

   third Solution: nuclear stain.

There is no need to wash with water between solutions, and the staining time for each of them will depend on the urgency of staining (*see* **Note 6**).

The advantages of this staining method are that once the cells are stained if further staining with Papanicolaou or another stain is needed, it can be done directly without the need for decolorization. It also does not require immediate mounting for visualization under the microscope (Figs. 7, 8 and 9).

When an immunohistochemical technique is required on a slide stained with Diff-Quik or Papanicolaou, remove this staining before following these steps:

- 10 min in xylene.
- 10 min in absolute alcohol.

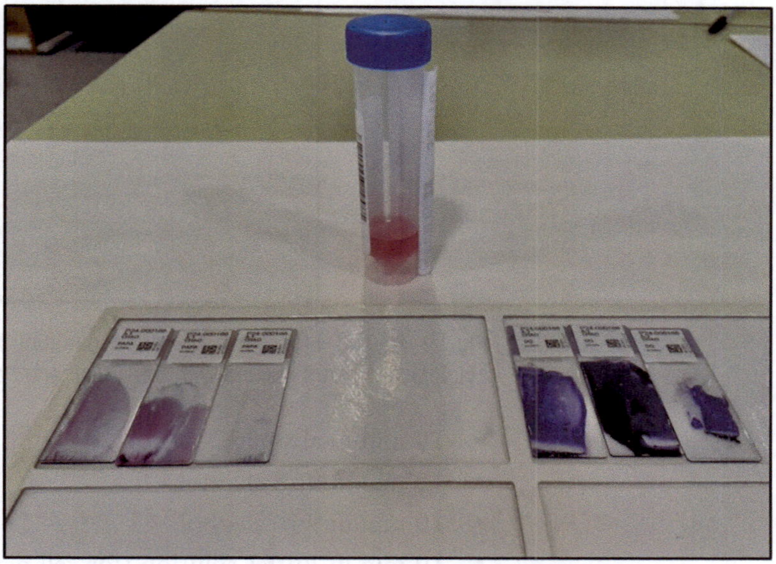

Fig. 7 Material preserved in liquid medium. After centrifugation, the supernatant is discarded and the cellular pellet is resuspended in PreservCyt

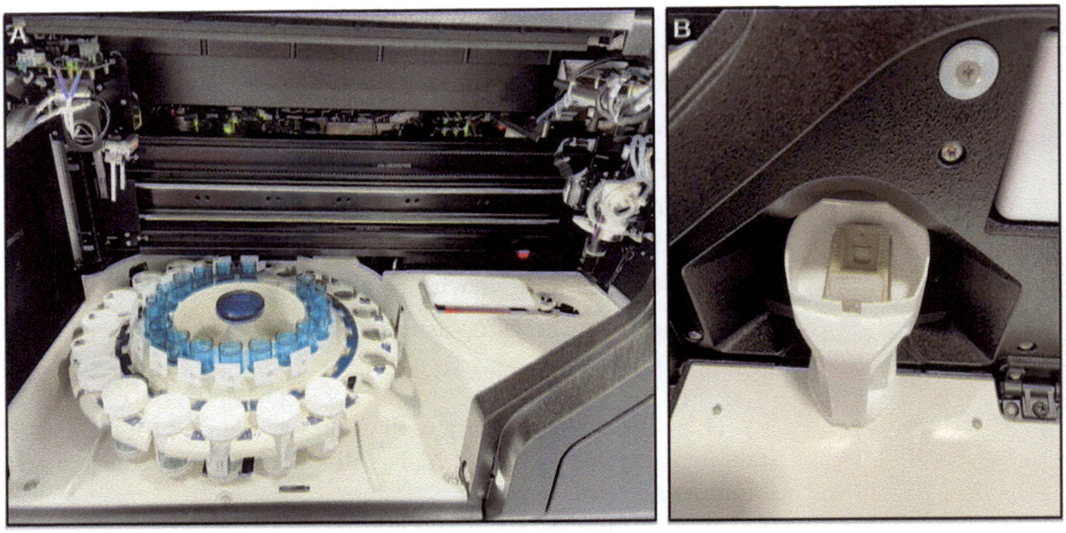

Fig. 8 (**a**) Automated ThinPrep processor. (**b**) Processed slides ready for subsequent staining

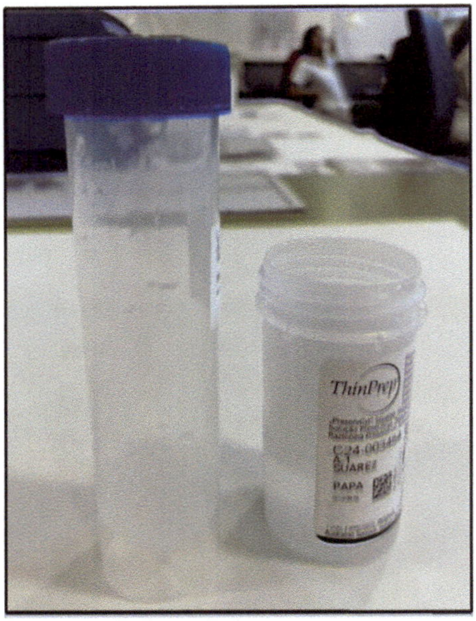

Fig. 9 Thyroid FNA (Fine Needle Aspiration) fluid in Cytolyt solution with cytological smears stained with Diff-Quick and Papanicolaou

- 10 min in absolute alcohol.
- 10 min in 96% alcohol.
- 10 min in 96% alcohol.
- 10 min in Buffer solution (for cell permeabilization and better results).

Slides processed automatically by ThinPrep (*see* **Note 7**) do not need to remove the previous staining; they are directly introduced into the last step of 10 min with the buffer solution.

- In both aforementioned cases (Subheading 2, **items 2.1** and **2.2**), they are processed in the automated immunostainer (steps described in Subheading 3, **step 1.3**). The protocol of the desired antibody is selected, and several distinguishable aspects are changed.
- Remove the Dewax solution from the protocol as it does not contain paraffin.
- Reduce the antigen retrieval time to a minimum (5 min) to prevent cell degradation.
- In the case that the desired antibody requires enzymatic digestion (very few cases), it needs to be replaced with the Epitope retrieval 1 the p solution.
- pH 6.

1.3 **Microscopic Examination:** Use light optical microscopy to identify any morphological abnormalities associated with disease or pathological conditions (*see* **Note 8**).

1.4 **Diagnostic Reporting:** Following microscopic analysis, a detailed pathological report must be generated, including a description of observed histological characteristics, results of immunohistochemical tests, and final diagnosis.

4 Notes

1. Each thyroid lobe is marked with different colored inks, or the anterior and posterior sides of the specimen are painted. This helps with spatial orientation and assessing margin involvement when viewing the specimen under the microscope.
2. We recommend that controls will be positioned at the top of the slide, while the sample under study will be placed at the bottom.
3. Standardized panels of different immunohistochemistry stains are available for the most prevalent pathologies. The ideal temperature for storing antibodies (concentrated or pre-diluted) and the detection kit (polymer) should be 4 °C (between 2 and 8 degrees), as freezing and thawing have a detrimental effect on their performance. Antibodies should be used within less than 6 months and never past their expiration date.

4. The time of incubation is stablished in the protocol specific of each treading house.
5. We recommend centrifuging the cells at 1800 rpm to achieve optimal cell concentration while preserving cellular morphology, as per the manufacturer's recommendation.
6. We recommended staining time between 1 and 3 min.
7. The automatic processing of this machine (Thinprep 5000, Hologic) is based on the ThinPrep technique, which performs the following tasks:
   (a) Dispersion: The filter rotates inside the sample vial and creates currents within the liquid, which are intense enough to separate debris and disperse mucus, but gentle enough not to negatively alter the appearance of cells or cause their rupture.
   (b) Cell collection: A slight vacuum is created inside the filter to accumulate cells on the outer surface of the membrane in a monolayer.
   (c) Cell transfer: Once the cells are collected on the membrane, the filter is inverted and lightly pressed onto the slide, resulting in a uniform distribution of cells in a defined circular area.
8. We strongly recommend that a pathologist visualize the specimens and issue the report. The pathologist should watch first the routine stains and then immunohistochemical stains to archive the diagnoses. It is recommended to first examine the entire specimen at low magnification and then gradually increase the magnification to visualize all the sections.

References

1. Rindi G, Mete O, Uccella S et al (2022) Overview of the 2022 WHO classification of neuroendocrine neoplasms. Endocr Pathol 33(1): 115–154. https://doi.org/10.1007/s12022-022-09708-2. Epub 2022 Mar 16
2. Bocker W, Dralle H, Dorn G (1981) Thyroglobulina: an immunohistochemical marker in thyroid disease. In: DeLellis RA (ed) Diagnostic immunohistochemistry. Masson, New York, pp 37–60
3. Liu H, Lin F (2015) Application of immunohistochemistry in thyroid pathology. Arch Pathol Lab Med 139(1):67–82. https://doi.org/10.5858/arpa.2014-0056-RA
4. Nonaka D, Tang Y, Chirigoba L et al (2008) Diagnostic utility of thyroid transcription factors PAX8 and TTF1 (FoxE1) in thyroid epithelial neoplasm. Mod Pathol 21:192–200
5. Laury AR, Perets R, Paio H et al (2011) A comprehensive analysis of PAX8 expression in human epithelial tumors. Am J Surg Pathol 35(6):816–826
6. Tacha D, Zhou D, Cheng L (2011) Expression of PAX 8 in normal and neoplastic tissues: a comprehensive immunohistochemical study. Appl Immunohistochem Mol Morphl 19(4): 293–299
7. Savin S, Cvejic D, IsicT et al (2008). Thyroid peroxidase and galectin-3 imunostaining in differentiated thyroid carcinoma with clinicopathologic correlation. Hum Pathol; 39(11): 1656–1663
8. Findlay DM, SextonPM (2004). Calcitonin: Growth Factors 22:217–24. Doi: https://doi.org/10.1080/08977190410001728033

9. Keen CE, Szakacs S, Okon E et al (1999) CA125 and thyroglobulin staining in papillary carcinoma of thyroid and ovarian origen is not enterely specific for site of origen. Histopathology 34:113–117
10. Ordoñez NG, El-Naggar AK, Hickey et al (1991). Anaplastic thyroid carcinoma: immunocytochemical study of 32 cases. Am J Clin Pathol; 96:15–24
11. Ordoñez NG (2000) Thyroid transcription factor1 is a marker of lung and thyroid carcinomas. Adv Anat Pathol 7:123–576
12. Miettinen M, Franssila K, Letho V et al (1984) Expression of intermediate filament proteins in thyroid gland and thyroid tumors. Lab Investig 50:262–227
13. Prasad ML, Pellegata NS, Huang Y et al (2005) Galectina-3, fibronectina-1, CITED1, HMBE1 and cytokeratin-19 immunohistochemistry is useful for the differential diagnosis of thyroid tumors. Mod Pathol 18:48–57
14. Mase T, Funahashi H, Koshikawa T et al (2003) HMBE1 immunostaining in thyroid tumors specially in follicular neoplasm. Endocr Pathol 14:55–60
15. Bartolazzi A, Gasbarri A, Papotti M et al (2001) Applicationof an immunodiagnostic method for improving pre-operative diagnosis of nodular thyroid lesions. Lancet 357:1644–1650
16. Fornaro M, Dell Arciprete R, Stella M et al (1995) Cloning of the gene encoding Trop-2, a cell-surface glycoprotein expressed by human carcinomas. Int J Cancer 62(5):610–618
17. Suzuki A, Hirokawa M, Takada N et al J (2018). Utility of monoclonal PAX 8 antibody for distinguishing intrathyroid thymic carcinoma from follicular cell-derived thyroid carcinoma. Endocr 65: 1171–1175. Doi: https://doi.org/10.1507/endocrj.EJ18-0282
18. Bishop JA, Sharma R, Westra WH (2011) PAX 8 immunstaining of anaplastic thyroid carcinoma: a realible means of discerning thyroid origin for undifferentiated tumors of the head and neck. Human Pathol 42:1873–1877. https://doi.org/10.1016/j.humpath.2011.02.004
19. Prasad ML, Pellegata NS, Kloos RT et al (2004) CITED1 expression suggests papillary thyroid carcinoma in highthroughput tissue microarray-based study. Thyroid 14:169–175
20. Hirokawa M, Carney JA (2000) Cell membrane and cytoplasmic staining for MIB-1 in hyalinizing trabecular adenoma of thyroid. Am J Surg Pathol 24:575–578
21. Smith AL, Willians MD, Stewart J et al (2020) Utility of the BRAF p V600E immunoperoxidase stai in FNA direct smears and cell block preparations from patients with solid tumours with microsatellite instability or NTRK fusions. Ann Oncol 31:861–872. https://doi.org/10.1016/j.annonc.2020.03299
22. Haugen BR, Alexander EK, Bible KC, Doherty GM, Mandel SJ, Nikiforov YE et al (2016) American Thyroid Association management guidelines for adult patients with thyroid nodules and differentiated thyroid cancer: the American Thyroid Association guidelines task force on thyroid nodules and differentiated thyroid cancer. Thyroid 26:1–133
23. Cameselle-Teijeiroa, J. M., Bella Cuetob, M. R., Eloyc et al (2020). Tumores de la glándula tiroides: Propuesta para el manejo y estudio de las muestras de pacientes con neoplasias tiroideas. Rev Esp Patol2020; 53 (1): 27–36. DOI: https://doi.org/10.1016/j.patol.2019.03.003

Chapter 16

Thyroid Hormones Determination in Euthyroid and LT4-Treated Patients During COVID-19 Hospitalization

Cristina Álvarez Castilla

Abstract

The global impact of the novel coronavirus disease (COVID-19), caused by the severe acute respiratory syndrome coronavirus 2 (SARS-CoV-2), was significant, including in Spain. The initial outbreak in late January 2020 prompted a frantic effort to comprehend the virus and contain its spread, given the limited knowledge available at the time. The study by Amich et al. (2022) in Frontiers in Endocrinology aimed to investigate the influence of thyroid hormone levels and age on the severity of COVID-19 in euthyroid and levothyroxine-treated patients. Although a direct correlation between thyroid hormone levels and the severity of COVID-19 has not been established, it is known that thyroid dysfunction affects immune function and general health, which may impact the outcomes of COVID-19. The research included patients from the Clinic San Carlos Hospital in Madrid, and a comprehensive range of data was collected from each patient, including demographic information, symptoms, comorbidities, and laboratory test results. This approach was employed in order to accurately match patients and identify factors that may influence the severity of COVID-19 outcomes. Blood samples were analyzed using chemiluminescence on a DXI-800® instrument, with the following hormones quantified: thyroid-stimulating hormone (TSH), free thyroxine (FT4), and free triiodothyronine (FT3). These insights contribute to an understanding of the interplay between thyroid function and the severity of COVID-19, highlighting the need for careful patient management in those with thyroid disorders during the pandemic.

Key words COVID-19 severity, Thyroid hormones, Thyroid-stimulating hormone, Comorbidities, Levothyroxine treatment, Chemiluminescence

1 Introduction

The onset of the COVID-19 pandemic in Spain, caused by the severe acute respiratory syndrome coronavirus 2 (SARS-CoV-2), as in numerous other regions around the globe, was marked by uncertainty and rapid adaptation. The pandemic's origins can be traced back to late January 2020, when the first imported cases were identified in Spain. The initial stages of the pandemic were characterized by a sense of urgency to comprehend the virus and its transmission dynamics, as well as to implement protocols to contain

Constanza Contreras-Jurado (ed.), *Thyroid Hormones: Methods and Protocols*, Methods in Molecular Biology, vol. 2876, https://doi.org/10.1007/978-1-0716-4252-8_16,

its spread. At the time, the knowledge of the virus was limited, and the situation was evolving rapidly.

The objective of the article "Free triiodothyronine levels and age influence the metabolic profile and COVID-19 severity parameters in euthyroid and levothyroxine-treated patients" by Amich et al. (2022), published in Frontiers in endocrinology, was to identify whether thyroid hormone levels and age were significant factors in the development of severe COVID-19 disease [1].

There is no established direct correlation between thyroid hormone levels and the severity of COVID-19 disease. However, it is well documented that thyroid dysfunction can impact immune function and general health, suggesting a potential influence on the course of a COVID-19 infection. Additionally, thyroid hormones play a role in regulating the immune system and inflammation [2]. The severity of COVID-19 is often associated with an exaggerated immune response leading to inflammation, which can contribute to complications such as acute respiratory distress syndrome (ARDS) [3].

The SARS-CoV-2 virus is associated with a respiratory infection characterized by pneumonia-like symptoms. These symptoms can vary in severity, from being completely asymptomatic to causing severe respiratory failure [4]. It is well established that thyroid dysfunction frequently coexists with other conditions, including cardiovascular disease, diabetes, and obesity. These conditions are well-known risk factors for severe COVID-19. The presence of these comorbidities may contribute to an increased risk of complications from COVID-19, rather than thyroid hormone levels per se [4, 5].

1.1 Patients' Selection

The samples utilized in the development of this article were provided by the biobank of the Clinic San Carlos Hospital in Madrid, with the ethical approval of the institution. The following groups were selected for analysis:

- Group 1: normothyroid patients infected with COVID-19, confirmed by polymerase chain reaction (PCR) or serology.
- Group 2: patients with a diagnosis of hypothyroidism who were infected with COVID-19 and confirmed by PCR or serology. These patients were admitted to a non-critical ward and did not require assisted ventilation.
- Group 3: patients with a diagnosis of hypothyroidism and confirmed COVID-19 infection (by PCR or serology). The patients were critically ill and required assisted ventilation, and were therefore admitted to the ICU.

Individuals with malignant neoplasms, transplant recipients, pregnant women, hyperthyroidism, or thyroid disorders that had not been confirmed during hospitalization, those prescribed less than 525 mg of T4 per week, and those below the age of 18 were excluded from the study [1].

In order to ascertain the potential role of thyroid hormones in the development of severe COVID-19 and the influence of other factors, such as comorbidities, it is crucial to match patients by these factors in order to ensure accurate results. The main purpose of matching is to increase the efficiency of the study in terms of data collection and subsequent statistical analysis [6]. Prior to initiating the study, it is essential to collect a comprehensive set of clinical data points pertaining to each individual participating in the study:

- General patient data: age, sex, days of hospitalization, days in the ICU, smoking status.
- Symptoms at admission: temperature, respiratory rate, heart rate, systolic blood pressure, diastolic blood pressure, symptoms of partial loss of consciousness, disorientation, fatigue, fever, headache, myalgia, lymphadenopathy, hemoptysis, conjunctivitis, nasal congestion, cough, nausea, vomiting, odynophagia, oropharyngeal congestion, chills, congestion, unilateral or bilateral pneumonia.
- Comorbidities: cardiovascular, cerebrovascular, diabetes, chronic obstructive pulmonary disease (COPD), hypertension, immunodeficiency, liver failure, neoplasia, renal failure.
- Blood tests: albumin, alanine aminotransferase (ALT), aspartate aminotransferase (AST), bilirubin, C-reactive protein, creatinine, creatine phosphokinase, D-dimer, ferritin, glucose, interleukin-6, lactate dehydrogenase, leukocytes, lymphocytes, monocytes, neutrophils, platelets, procalcitonin, proteins, prothrombin, urea.

The data collected is used to identify potential comorbidities, which assist in matching patients. Blood tests are used as parameters to indicate the severity of the infection [1]. The current evidence suggests that a number of factors may increase the risk of developing severe COVID-19 in adults. These include older age, male sex, ethnicity, and the presence of underlying diseases. A higher risk for a fatal course of COVID-19 was observed in individuals with obesity, diabetes, chronic respiratory disease, cardiovascular disease, and immunosuppressive conditions such as cancer [4, 5, 7, 8].

1.2 Risk Factors Selected to Be Included in the Study

1.2.1 Age

Aging has been demonstrated to be a significant risk factor for the development of severe COVID-19. Individuals of advanced age frequently present with multiple concurrent health conditions, or comorbidities. Furthermore, a weaker immune response to infectious agents and an age-related chronic pro-inflammatory status of the immune system, accompanied by persistent low-grade innate immune activation and an elevated level of pro-inflammatory cytokines, may contribute to an increased susceptibility to tissue damage caused by infections in the elderly [8].

1.2.2 Gender

Several studies indicate a higher susceptibility of male individuals to severe infections due to the SARS-CoV-2 virus. This may be due to the effects of different hormones on the inflammatory process, differences in angiotensin receptor (ACE2) levels and environmental factors such as lifestyle [8].

1.2.3 Obesity

Obesity, defined as a body mass index (BMI) ≥ 30 kg/m^2, is a risk factor for an increased severity of COVID-19 disease [4, 5]. Obesity has been described as a risk factor for the development of comorbidities, which may themselves lead to severe COVID-19 disease. However, obesity also presents as a risk factor for the development of viral infections. This is due to a reduction in oxygen saturation of the blood, resulting from compromised ventilation at the base of the lungs and the production of abnormal levels of cytokines and interferon. This may subsequently compromise the immune response, and produces an underlying chronic inflammatory phenotype [5, 7].

1.2.4 Diabetes Mellitus

Diabetes is associated with an increased risk of severe outcomes in patients infected with SARS-CoV-2, which include acute respiratory failure syndrome, pneumonia, multiple organ failure, or death [7]. Diabetes and hyperglycemia, in particular, is known to be associated with immune dysfunction such as cytokine dysregulation, excessive protein glycation, and increased oxidative stress leading to a pro-inflammatory state [4, 5, 7]. In addition, coronavirus infection has been shown to damage pancreatic islet cells and cause acute insulin-dependent diabetes, increasing the patient's vulnerability [5, 7].

1.2.5 Chronic Respiratory Disease

Individuals with chronic respiratory diseases such as chronic obstructive pulmonary disease (COPD) or asthma are at an increased risk of developing severe outcomes associated with the COVID-19 disease [4].

Chronic Obstructive Pulmonary Disease (COPD)

The expression of ACE-2 receptors is increased in this disease. This contributes to the establishment of severe symptoms of infection, including structural damage to the lungs, weakened immunity, and hypermucus production [4].

Asthma

People with asthma have a delayed innate antiviral immune response and impaired interferon secretion, making them more susceptible to developing severe complications of SARS-CoV-2 infection [4].

1.2.6 Cardiovascular Disease

Hypertension

Hypertension is one of the most prevalent underlying comorbidities of COVID-19 disease. However, the extent to which hypertension itself contributes to the development and progression of the disease remains uncertain. It is likely that the association of hypertension with advanced age and the coexistence of other comorbidities may contribute to a compromised immune response, potentially exacerbating the severity of COVID-19 symptoms [5, 7].

Cardiovascular Disease (CVD)

Cardiovascular disease (CVD) is frequently linked to the development of severe COVID-19. Although the precise mechanism underlying this phenomenon is not yet fully understood, it is known that these patients typically have a compromised immune system. One proposed mechanism of pathogenesis is the expression of the ACE-2 receptor on cardiac muscle cells, leading to direct infection of the myocytes. Additionally, the cytokine storm initiated by the immune system during the infection's development contributes to ischemia and thrombosis [4, 7].

1.2.7 Immunocompromised States: Cancer

In cancer, the immune system is compromised by the tumor itself and by numerous anticancer treatments, increasing the risk of infection and exacerbating symptoms, including those associated with SARS-CoV-2 [4, 7].

2 Materials

Blood samples were collected, serum was extracted, and aliquots were placed in sterile 1.5-ml tubes, which were subsequently stored at a temperature of −80 °C until required.

The thyroid-stimulating hormone (TSH), free thyroxine (FT4), and free triiodothyronine (FT3) were evaluated by chemiluminescence on a DXI-800® (Beckman-Coulter®). These measurements were conducted in accordance with the manufacturer's instructions using three distinct assays and original reagents [9, 10]:

- Access TSH (third IS) assay.

The reagents are stable for a period of 63 days after initial use when stored at temperatures of 2 and 10 °C.

1. R1a

Paramagnetic particles coated with mouse monoclonal anti-human TSH antibody suspended in TRIS (Tris(hydroxymethyl)-aminomethane) buffered saline at pH 7.4, with surfactant, bovine serum albumin (BSA), and sodium azide (NaN_3) at a concentration of less than 0.1%.

2. R1b

TRIS (Tris(hydroxymethyl) aminomethane) buffered saline at pH 7.4 with surfactant, BSA, protein (murine), and sodium azide (NaN_3) at a concentration of less than 0.1%.

3. R1c and R1d

Mouse monoclonal anti-human TSH alkaline phosphatase conjugate in ACES (N-(2-Acetamido)-2-aminoethanesulfonic acid) buffered saline at pH 6.0, with surfactant, BSA matrix, protein (murine), and sodium azide (NaN_3) at a concentration of less than 0.1%.

- Access Free T4 assay.

  The reagents are stable for a period of 28 days after initial use when stored at temperatures of 2 and 10 °C.

1. R1a

Paramagnetic particles coated with streptavidin in a TRIS (Tris (hydroxymethyl)aminomethane) buffer at pH 8.0 with protein (aves), surfactant, and sodium azide (NaN_3) at a concentration of 0.125%.

2. R1b

TRIS buffered saline (pH 8.0) with protein (aves), surfactant, and $< 0.1\%$ NaN_3.

3. R1c

TRIS (Tris(hydroxymethyl)aminomethane) buffered saline at pH 8.0 with protein (aves), surfactant, and sodium azide (NaN_3) at a concentration of 0.125%.

4. R1d

Triiodothyronine-alkaline phosphatase (bovine) conjugate in a TRIS (Tris(hydroxymethyl)aminomethane) buffer at pH 7.0 with protein (aves), surfactant, and sodium azide (NaN_3) at a concentration of less than 0.1%.

5. R1e

Mouse monoclonal anti-Thyroxine (T4) coupled to biotin in a TRIS (Tris(hydroxymethyl)aminomethane) buffer at pH 8.0 with protein (aves and murine), surfactant, and sodium azide (NaN_3) at a concentration of 0.125%.

- Access Free T3 assay.

  The reagents are stable for a period of 28 days after initial use when stored at temperatures of 2 and 10 °C.

1. R1a

Paramagnetic particles coated with streptavidin in a TRIS (Tris (hydroxymethyl)aminomethane) buffer at pH 8.0 with protein (aves), surfactant, and sodium azide (NaN_3) at a concentration of less than 0.1%.

2. R1b

MES (2-(N-morpholino) ethanesulfonic acid) buffer at pH 5.5.

3. R1c

Biotinylated T3 analog in a TRIS (Tris(hydroxymethyl)-aminomethane) buffer at pH 8.0 with protein (aves), surfactant, and sodium azide (NaN_3) at a concentration of less than 0.1%.

4. R1d

TRIS (Tris(hydroxymethyl)aminomethane) buffer at pH 8.0 containing animal protein (goat, bovine, aves), surfactant, and sodium azide (NaN_3) at a concentration of less than 0.1%.

5. R1e

Monoclonal antibody-alkaline phosphatase in an ACES (N-(2-Acetamido)-2-aminoethanesulfonic acid) buffer at pH 7.0 with protein (aves), surfactant, and sodium azide (NaN_3) at a concentration of less than 0.1%.

- Access substrate Lumi-Phos* 530.

  Is suitable for chemiluminescent detection of phosphatase alkaline conjugates in immunoassays. The co-surfactant system promotes and enhances energy transfer to the fluorescent head groups, resulting in luminescence with a broad emission maximum at 530 nm [11].

3 Methods

Chemiluminescence is a biochemical phenomenon that results in emission of light as a consequence of a chemical reaction, exhibiting varying degrees of intensity and lifetime, with colors that span the visible spectrum. This phenomenon has a variety of applications in scientific research, diagnostics, and other fields [12].

In the DXI-800® (Beckman-Coulter®), the reactions occur within the instrument. In the context of thyroid hormone measurement, the chemiluminescence reaction employs Lumi-Phos* 530 as a substrate. During this reaction, Lumi-Phos* 530 is dephosphorylated by the enzyme alkaline phosphatase (AP), which serves as a catalyst for the reaction, resulting in the production of light. This light can be detected by a luminometer integrated into the instrument [9–11].

3.1 TSH Measurement

The Access TSH (third IS) assay is a paramagnetic particle, chemiluminescent immunoassay for the quantitative determination of human thyroid-stimulating hormone (TSH) levels in human serum and plasma using the Access Immunoassay Systems. In terms of methodology, the Access TSH (third IS) assay is a two-site immunoenzymatic assay or two-site sandwich assay, comprising the following phases [9, 10] (Fig. 1):

1. Firstly, the serum sample is introduced into the apparatus and then pipetted into a reaction vessel together with the following reagents:
   (a) R1c and R1d: Mouse anti-TSH-alkaline phosphatase conjugate. The alkaline phosphatase is responsible for the chemiluminescence reaction.
   (b) R1b: Buffered protein solution.
   (c) R1a: Paramagnetic particles coated with immobilized mouse monoclonal anti-TSH antibody, which are called solid phase.
2. During the incubation period, the free TSH present in the patient's sample will bind to both types of antibodies. These antibodies will then react with a different antigenic site on the TSH, creating a sandwich-like structure that gives the technique its name.

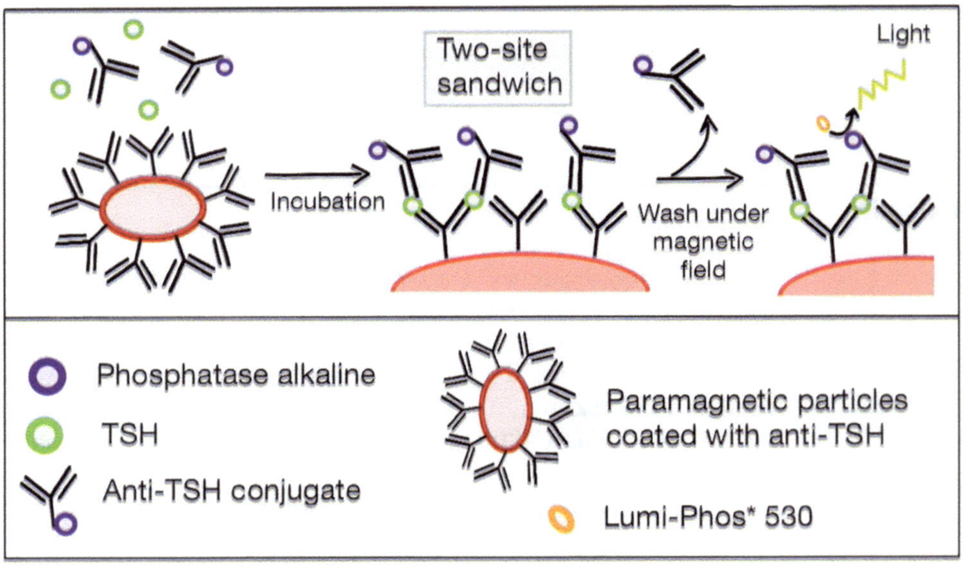

Fig. 1 A schematic representation of the two-site immunoenzymatic or two-site sandwich assay reaction for the measurement of TSH in serum using the Beckman Coulter DXI800® kit

3. Subsequently, the apparatus applies a magnetic field, whereby the solid phase is retained within the reaction vessel, while unbound components are washed away.
4. Following this washing, the chemiluminescent substrate is added to the reaction vessel (Lumi-Phos* 530), and the light generated by the reaction is measured by a luminometer.
5. The light production is directly proportional to the concentration of TSH in the sample. The amount of analyte in the sample is determined from a stored, multi-point calibration curve.

3.2 Free Thyroxine (FT4) Measurement

The Access Free T4 assay is a paramagnetic particle, chemiluminescent immunoassay for the quantitative determination of free thyroxine levels in human serum and plasma (heparin) using the Access Immunoassay Systems. The Access Free T4 assay is a two-step enzyme immunoassay comprising the following phases [9, 10] (Fig. 2):

1. As in the previous methodology, the serum sample is first introduced into the apparatus and then pipetted into a reaction vessel together with the following reagents:
   (a) R1e: Anti-Thyroxine (T4) antibody coupled to biotin.
   (b) R1c: Buffered protein solution.
   (c) R1a: Streptavidin-coated paramagnetic particles, which constitute the solid phase.
2. During the initial incubation period, the unbound thyroxine present in the sample will bind to the anti-T4 antibody, which will subsequently join the solid phase through the streptavidin-biotin interaction.

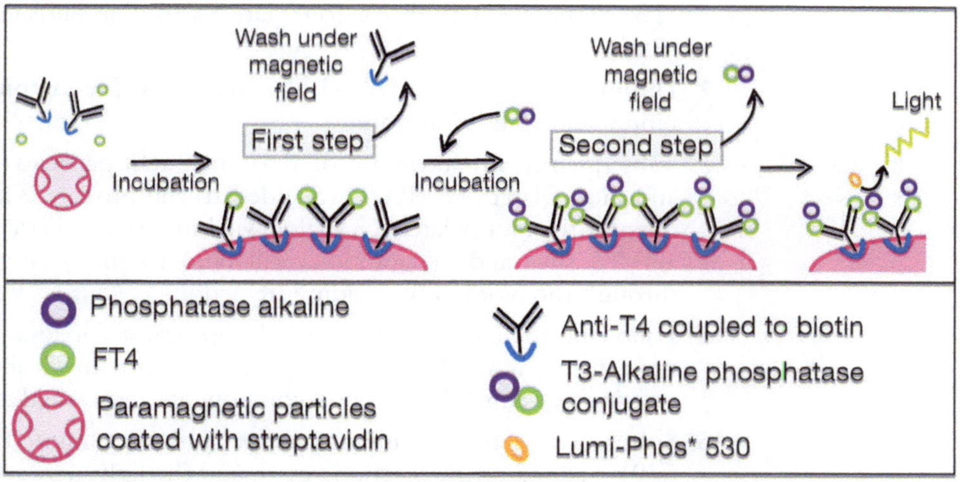

Fig. 2 A schematic representation of a two-step enzyme immunoassay reaction for the measurement of FT4 in serum using the Beckman Coulter DXI800® kit

3. Following this incubation period, the equipment applies a magnetic field in which the solid phase is retained within the reaction vessel, while any unbound components are washed away.
4. The subsequent step involves the addition of a buffered protein solution (R1b) and a triiodothyronine (T3)-alkaline phosphatase conjugate (R1d) to the reaction vessel. During this incubation period, the T3-alkaline phosphatase conjugate binds to the vacant anti-T4 antibody binding sites.
5. Subsequently, another washing process under a magnetic field is required to eliminate all components that have not bound to the solid phase.
6. Then, the chemiluminescent substrate (Lumi Phos* 530) is added to the reaction vessel and the light generated is measured with a luminometer.
7. The light production is inversely proportional to the concentration of free T4 in the sample. The amount of analyte in the sample is determined from a stored, multi-point calibration curve.

3.3 Free Triiodothyronine (FT3) Measurement

The Access Free T3 assay is a paramagnetic particle, chemiluminescent immunoassay for the quantitative determination of free triiodothyronine levels in human serum and plasma using the Access Immunoassay Systems. The Access Free T3 assay is a competitive binding immunoenzymatic assay, comprising the following phases [9, 10] (Fig. 3):

1. As in the previous methodologies, the serum sample is first introduced into the apparatus and then pipetted into a reaction vessel, where it is combined with an anti-triiodothyronine (T3) monoclonal antibody conjugate to alkaline phosphatase (R1e).
2. During the incubation period, the free T3 in the sample will bind to the anti-T3 antibody.
3. Subsequently, streptavidin-coated paramagnetic particles (R1a) and biotinylated T3 (R1c) are added to the mixture. Consequently, the biotinylated T3 analog will bind to the unoccupied anti-T3 sites and, in addition, will bind to the solid phase through the biotin-streptavidin interaction.
4. Following the incubation period, the apparatus applies a magnetic field in which the solid phase is retained within the reaction vessel, while unbound components are washed away.
6. Subsequently, the chemiluminescent substrate (Lumi Phos* 530) is added to the reaction vessel, and the light generated is measured with a luminometer.

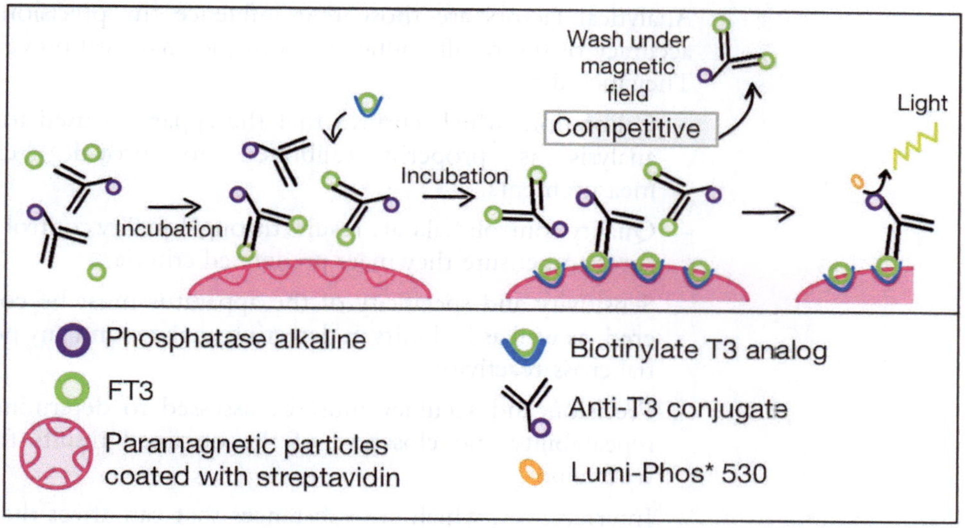

Fig. 3 A schematic representation of a competitive binding immunoenzymatic assay reaction for the measurement of FT3 in serum using the Beckman Coulter DXI800® kit

7. The light production is inversely proportional to the concentration of free T3 in the sample. The amount of analyte in the sample is determined from a stored, multi-point calibration curve.

3.4 Results Interpretation

The interpretation of results following the processing of samples in an apparatus necessitates an understanding of several key factors in order to ensure accuracy and reliability.

- Preanalytical factors are of paramount importance in any laboratory analysis. These include the manner of sample collection, the procedures used for sample preparation prior to analysis, and the conditions of transportation and storage. Each of these factors can influence the integrity of the sample and, consequently, the accuracy of the analysis.
  - Sample Collection: It is of the utmost importance to ensure that the correct collection techniques are employed in order to prevent contamination or degradation (*see* **Note 1**).
  - Sample Preparation: The preparation of samples is a crucial step in any analytical process. It is imperative to adhere to the prescribed procedures for the analysis of samples to prevent errors (*see* **Note 2**).
  - Transport and Storage: It is essential to maintain optimal conditions throughout the transportation and storage phases to ensure the integrity of the sample is preserved (*see* **Notes 3** and **4**).

- Analytical factors are those that influence the precision and accuracy of the results obtained from the analytical procedure. They include:
  - Calibration, which ensures that the apparatus used for the analysis is properly calibrated to provide reliable measurements.
  - Quality control: Validate results through quality control measures and ensure they meet predefined criteria.
  - Sensitivity and specificity of the apparatus must be considered, as well as its limits in detecting analytes and any potential cross-reactivity.
  - Precision and accuracy must be assessed to determine the repeatability and closeness of the obtained results to the true value.
  - Interferences, which are substances that can affect the outcome of the analysis.

The principal interferences in these assays include:

Antibody-Mediated Interference (*See* Note 5) The antibodies present in the patient serum may interact with either the analyte or the antibodies utilized in the assay. This interaction is particularly evident in sandwich immunoassays, but it has also been observed in competitive immunoassays. This can lead to erroneous results in the quantification of analytes [13, 14], as illustrated in the examples presented in Fig. 4:

- *Figure 4a: In the context of competitive immunoassays for the measurement of FT3, an interaction can occur between the analyte and antibodies present in the serum, resulting in inhibition of the reaction between the analyte molecules and the detection antibodies. In this example, detection antibodies that have become free due to the presence of antibodies in the serum can then bind to the biotinylated FT3 analog that is retained to the solid phase, resulting in an increase in the light produced. This phenomenon is known as "negative interference" and can result in a false reduction in the quantification of FT3.*
- *Figure 4b: In the sandwich TSH assay, positive interference occurs when interfering antibodies bind to detecting antibodies present in the assay by cross-linking in the absence of antigen, resulting in an overestimated result*

Biotin Interference (*See* Note 6) The recommended daily dose of biotin to treat hair loss, nail growth, and/or skin problems is 30 μg [15]. This dose is not likely to cause significant interference because serum biotin levels would be <10 ng/mL [13]. However, biotin in

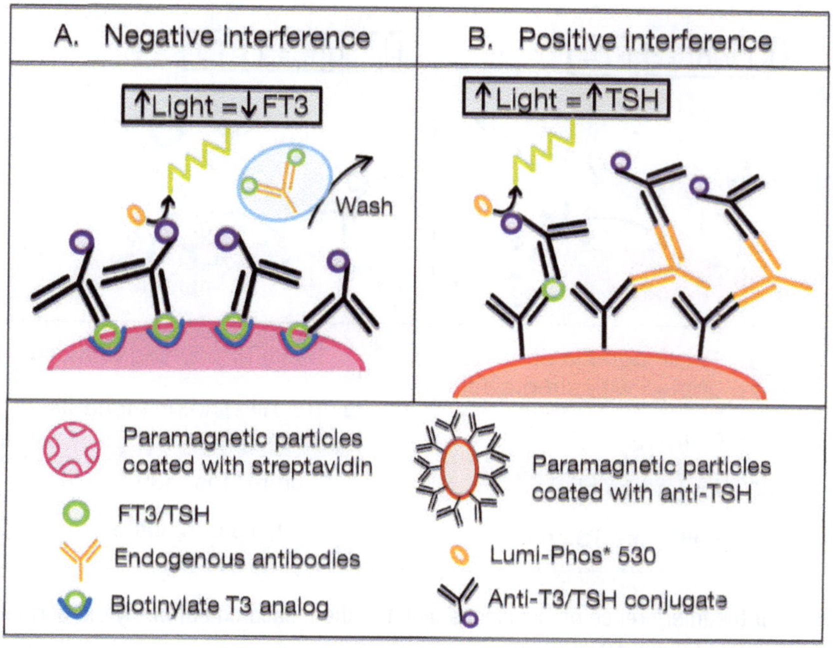

Fig. 4 Illustration of negative (**a**) and positive (**b**) interference in the measurement of thyroid hormones due to the presence of antibodies in the sample

high doses (ranging between 100 and 300 milligrams per day) is used in novel therapeutic approaches against multiple sclerosis. These doses result in serum concentrations >10 ng/mL, thereby leading to significant interference in the measurement of free thyroxine and free triiodothyronine by immunoassays employing the streptavidin–biotin method [13]:

> The presence of biotin in the patient's serum will displace the assay biotin from its binding sites on streptavidin, thereby inhibiting its function during the immunoassay. This phenomenon ultimately results in the washout of alkaline phosphatase and the subsequent reduction of light intensity in assays. Given that the light intensity produced is inversely proportional to the amount of analyte in these two assays, the interference results in an overestimation of the concentration of FT3 and FT4 in the sample (Fig. 5).

4 Notes

When measuring thyroid hormones on a Beckman Coulter DXI800® analyzer, it is important to consider several potential interferences and factors in order to ensure accurate and reliable results [9]:

1. Hemolysis and lipemia may potentially interfere with the TSH assay, resulting in erroneous results [9].

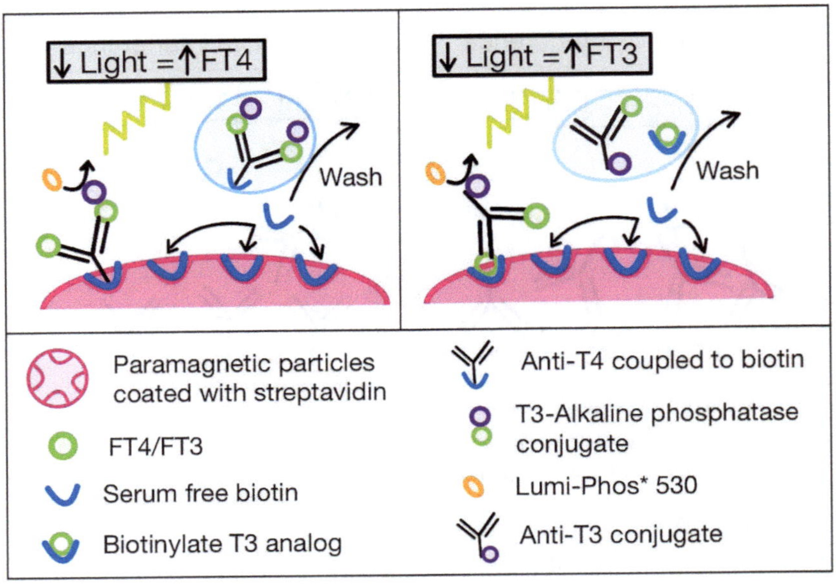

Fig. 5 Illustration of the interference of exogenous biotin in the measurement of thyroid hormones (FT3 and FT4) using the Beckman Coulter DXI800® kit

2. It is recommended that serum samples be allowed to clot completely prior to centrifugation in order to ensure that any residual fibrin and cellular matter has been removed prior to analysis [9].
3. For FT4 and FT3 measurement it is recommended that samples be stored in a stoppered container at room temperature (15–30 °C) in order to prevent deterioration for a maximum period of 8 h. For longer periods, samples should be refrigerated at 2–8 °C until a maximum of 48 h. If the assay will not be completed within this time frame, or if samples are to be shipped, they should be frozen at −20 °C or colder. In the case of the FT3 measurement, a maximum of three thaws is permitted, in contrast to the FT4, which is permitted a maximum of one thaw [9].
4. For TSH measurement, samples should be stored at room temperature (15–30 °C) in a covered container for up to 18 h to prevent deterioration. For longer periods, samples should be refrigerated at 2–8 °C for up to 7 days. In the event that the assay is not to be completed within the specified timeframe, the samples should be frozen at a temperature of −20 °C or below for a maximum of 90 days, with a maximum of two thaws permitted [9].
5. The presence of heterophile antibodies in serum samples may potentially interfere with the results of the assays. Additionally, patients who have received immunotherapy or diagnostic

procedures utilizing immunoglobulins may produce antibodies that interfere with immunoassays. In addition, interference has been observed in cases of high levels of rheumatoid factor (IgM type autoantibody) [9].

6. The assays employed to quantify FT3 and FT4 are vulnerable to the effects of high doses of biotin, which can give rise to significant interferences [9].

References

1. Amich I, Anguita E, Escribano-Serrat S et al (2022) Free triiodothyronine levels and age influences the metabolic profile and COVID-19 severity parameters in euthyroid and levothyroxine-treated patients. Front Endocrinol (Lausanne) 13:1025032. https://doi.org/10.3389/fendo.2022.1025032
2. van der Spek AH, Fliers E, Boelen A (2017) Thyroid hormone metabolism in innate immune cells. J Endocrinol 232(2):R67–R81. https://doi.org/10.1530/JOE-16-0462
3. Azkur AK, Akdis M, Azkur D et al (2020) Immune response to SARS-CoV-2 and mechanisms of immunopathological changes in COVID-19. Allergy 75(7):1564–1581. https://doi.org/10.1111/all.14364
4. Ejaz H, Alsrhani A, Zafar A et al (2020) COVID-19 and comorbidities: deleterious impact on infected patients. J Infect Public Health 13(12):1833–1839. https://doi.org/10.1016/j.jiph.2020.07.014
5. Müller-Wieland D, Marx N, Dreher M, Fritzen K, Schnell O (2022) COVID-19 and cardiovascular comorbidities. Exp Clin Endocrinol Diabetes 130(3):178–189. https://doi.org/10.1055/a-1269-1405
6. Iwagami M, Shinozaki T (2022) Introduction to matching in case-control and cohort studies. Ann Clin Epidemiol 4(2):33–40. https://doi.org/10.37737/ace.22005
7. Silaghi-Dumitrescu R, Patrascu I, Lehene M, Bercea I (2023) Comorbidities of COVID-19 patients. Medicina (Kaunas) 59(8):1393. https://doi.org/10.3390/medicina59081393
8. Zhang JJ, Dong X, Liu GH, Gao YD (2023) Risk and protective factors for COVID-19 morbidity, severity, and mortality. Clin Rev Allergy Immunol 64(1):90–107. https://doi.org/10.1007/s12016-022-08921-5
9. Beckman Coulter Company, Instructions for use. https://www.beckmancoulter.com/en/products/immunoassay/dxi-800?id=91501#/test-menu. Accessed 31 May 2024
10. Walsh M, Wener M. Laboratory procedure manual. National Health and Nutrition Examination Survey. 2001-2002 Data Documentation, Codebook, and Frequencies, https://wwwn.cdc.gov/Nchs/Nhanes/2001-2002/SSNH4THY.htm. Accessed 31 May 2024
11. Lumigen, A Beckman Coulter Company. http://www.lumigen.com/products/elisa/lumi-phos-530. Accessed 31 May 2024
12. Azim MAU, Hasan M, Ansari IH, Nasreen F (2015) Chemiluminescence immunoassay: basic mechanism and applications. Bangladesh J 18(2):171
13. Paczkowska K et al (2020) Laboratory interference in the thyroid function test. Endokrynol Pol 71(6):551–560. https://doi.org/10.5603/EP.a2020.0079
14. García-González E, Aramendía M, Álvarez-Ballano D, Trincado P, Rello L (2015) Serum sample containing endogenous antibodies interfering with multiple hormone immunoassays. Laboratory strategies to detect interference. Pract Lab Med 4:1–10. https://doi.org/10.1016/j.plabm.2015.11.001
15. National Institutes of Health, Office of dietary supplements. https://ods.od.nih.gov/factsheets/Biotin-HealthProfessional/. Accessed 9 June 2024

Correction to: Experimentally Manipulating the Thyroid Hormone Axis in Zebrafish

Melody Autumn, Jenny Zeng, Isabella Ranieri, and Sarah K. McMenamin

Correction to:
Chapter 13 in: Constanza Contreras-Jurado (ed.), *Thyroid Hormones: Methods and Protocols,* Methods in Molecular Biology, vol. 2876,
https://doi.org/10.1007/978-1-0716-4252-8_13

The original version of this chapter has the incorrect author name as 'Isabella Raineri'.

It has been corrected as 'Isabella Ranieri'.

The updated version of this chapter can be found at
https://doi.org/10.1007/978-1-0716-4252-8_13

Constanza Contreras-Jurado (ed.), *Thyroid Hormones: Methods and Protocols*, Methods in Molecular Biology, vol. 2876,
https://doi.org/10.1007/978-1-0716-4252-8_17,

Correction to: Thyroid Hormones

Constanza Contreras-Jurado

Correction to:
Constanza Contreras-Jurado (ed.), ***Thyroid Hormones: Methods and Protocols,*** **Methods in Molecular Biology, vol. 2876,**
https://doi.org/10.1007/978-1-0716-4252-8

- The first name of Constanza Contreras-Jurado was unfortunately published with an error. The initially published version has now been corrected. The correction chapters and the book have been updated with these changes.
- And the below third affiliation was added to Constanza Contreras-Jurado.
  "Centro de Investigación Biomédica en Red de Cáncer (CIBERONC), Instituto de Salud Carlos III, Madrid, Spain"

The updated version of this book can be found at
https://doi.org/10.1007/978-1-0716-4252-8

Constanza Contreras-Jurado (ed.), *Thyroid Hormones: Methods and Protocols*, Methods in Molecular Biology, vol. 2876,
https://doi.org/10.1007/978-1-0716-4252-8_18,

Index

Constanza Contreras-Jurado (ed.), *Thyroid Hormones: Methods and Protocols*, Methods in Molecular Biology, vol. 2876,
https://doi.org/10.1007/978-1-0716-4252-8,

MIX
Papier aus verantwortungsvollen Quellen
Paper from responsible sources
FSC® C105338

If you have any concerns about our products,
you can contact us on
ProductSafety@springernature.com

In case Publisher is established outside the EU,
the EU authorized representative is:
Springer Nature Customer Service Center GmbH
Europaplatz 3, 69115 Heidelberg, Germany

Printed by Libri Plureos GmbH
in Hamburg, Germany